iRODS
User Group Meeting 2019
Proceedings

11TH ANNUAL CONFERENCE SUMMARY

The iRODS User Group Meeting of 2019 gathered together iRODS users, Consortium members, and staff to discuss iRODS-enabled applications and discoveries, technologies developed around iRODS, and future development and sustainability of iRODS and the iRODS Consortium.

The four-day event was held from June 25th to 28th in Utrecht, Netherlands, hosted by Utrecht University and the iRODS Consortium, with over 130 people attending. Attendees and presenters represented over 60 academic, government, and commercial institutions.

TALKS AND PAPERS

LIGHTNING TALKS

Providing validated, templated and richer metadata using a bidirectional conversion between JSON and iRODS AVUs

J. Paul van Schayck
MUMC+ DataHub
P. Debyelaan 25,
Maastricht, The
Netherlands
p.vanschayck@maastrichtuniversity.nl

Ton Smeele
Utrecht University
ITS/RDM
Heidelberglaan 8,
Utrecht, The Netherlands
a.p.m.smeele@uu.nl

Daniël Theunissen
MUMC+ DataHub
P. Debyelaan 25,
Maastricht, The
Netherlands
d.theunissen@maastrichtuniversity.nl

Lazlo Westerhof
Utrecht University
ITS/RDM
Heidelberglaan 8,
Utrecht, The Netherlands
l.r.westerhof@uu.nl

ABSTRACT

A frequently recurring question in research data management is to structure metadata according to a standard and to provide the corresponding user interface to it. This has only become more urgent since the introduction of the FAIR principles which state that metadata should use controlled vocabularies and meet community standards.

The iRODS data grid technology is well positioned as a core layer within an infrastructure to manage research data. One of its strengths is the ability to attach any number of attribute, value, unit (AVU) triples as metadata to any iRODS object. This makes iRODS adaptable to very diverse use cases in research data management. However, the challenge of working with more structured metadata is not being addressed by the default capabilities of iRODS. Our aim is to develop a new method for storing richer, templated and validated metadata in AVUs.

JSON is a popular, flexible and easy to use format for serializing (nested) data, while maintaining human and developer readability. Furthermore, a JSON Schema can be used to validate a JSON structure and it can also be used to obtain a dynamically generated form on the basis of this schema. This combination of functionalities makes it an excellent format for metadata. We have therefore designed and implemented a bidirectional conversion between JSON and AVUs. The conversion method has been implemented as Python iRODS rules that allow to set and retrieve AVU metadata on an iRODS object using a JSON structure. Optionally, a policy can be installed to validate metadata entry and updates against the JSON Schema that governs the object.

With this work we provide other iRODS developers with a generic method for conversion between JSON and AVUs. We are encouraging others to use the conversion method in their deployments.

Keywords

Metadata, AVUs, JSON, conversion, validation, presentation

INTRODUCTION

Over recent years there has been an increasingly urgent call for the practice of open science [1]. Driven by the core values in research of transparency and reproducibility, the sharing of research data has become a hot topic.

Additional reasons for sharing research data are to better leverage investments in research by promoting reuse and making those data that can be considered public assets, available to the public [2]. To facilitate this accountability and reuse of research data, the Findable, Accessible, Interoperable and Reusable (FAIR) principles of research data have been introduced and broadly taken up [3]. Briefly, the FAIR principles suggest for research data to be globally and uniquely identifiable, and associated with searchable metadata ("Findable"); these identifiers should point to (meta)data using an open protocol ("Accessible") and that this data uses a formal representation language using widely applicable ontologies ("Interoperable"); finally, data should be provided with cross-references, provenance and license information ("Reusable"). Furthermore, the FAIR principles state that all this should be provided in both human and machine-readable form to facilitate automated pipelines and the increasing need for automated analysis and large-scale data research. However, the FAIR guidelines did not define any form of implementation recommendations for these principles.

The iRODS data grid technology is well-positioned as a core layer within an infrastructure to manage research data [4]. iRODS features support for preservation properties that are important for research data management such as authenticity, integrity and chain of custody. Most of these properties are met using metadata to annotate data objects and collections. Within iRODS, a basic building block for managing metadata is the AVU, short for *Attribute-Value-Unit*. The name "AVU" refers to its compound structure: It consists of three string-typed fields that as a triple represent a single metadata property of the data. While typically more than one AVU is needed to communicate and to document properties of research data[1], currently the iRODS support for AVU *composition* is somewhat limited. Microservices operate either on a single AVU or on an attribute-value map structure that does not allow a unit component to be specified. Attributes with multiple values, nested structures and atomic operations on a composition of AVUs are not supported. There is also no way to specify a template or structure that AVUs associated to an object should adhere to. These limitations may hinder the implementation of the FAIR principles or could lead to *ad hoc* solutions that are not transferable to other systems.

In contrast, many applications have adopted the data interchange standard JavaScript Object Notation (JSON) to efficiently exchange and operate on composed data structures [5]. JSON is lightweight and can be used across many programming languages. JSON data structures consist of an unordered collection of name/value pairs referred to as an *object*. Values are primitive data types, or an ordered list of values, or another JSON object. To improve interoperability, in 2017 IETF has published a more restrictive version of the JSON standard [6]. For instance, this version requires the name of name/value pairs to be unique, so that programming languages can conveniently implement JSON using map constructs.

JSON can be used to serialize arbitrary metadata, resulting in an equally arbitrary set of AVUs. For the purpose of adhering to the FAIR principles however, we seek to restrict the metadata that documents research data to a well-defined set of composed, related and consistent properties. The semantics of this metadata structure can be modeled as a *template*. Such a template can be applied to validate operations that attempt to modify any of the AVUs within the *namespace* of the template.

We propose to use JSON Schema as a technique to represent a metadata template within the context of iRODS. The metadata template will act on a composition of AVUs that is represented by a JSON data structure. JSON Schema is a draft internet standard that aims to define the structure and content of JSON formatted data [7]. Using a JSON Schema definition, applications such as iRODS can validate and interact with instances of JSON formatted data.

The application of a JSON Schema based template does not need to be limited to iRODS server-side validation. Figure 1 shows how client applications can opt to use the same JSON Schema in a presentation layer to *dynamically* render forms that facilitate data entry of metadata. For instance the React[2] component react-jsonschema-form[3] is

---

[1]see for instance the DataCite specification at https://schema.datacite.org/
[2]see https://reactjs.org
[3]see https://github.com/rjsf-team/react-jsonschema-form

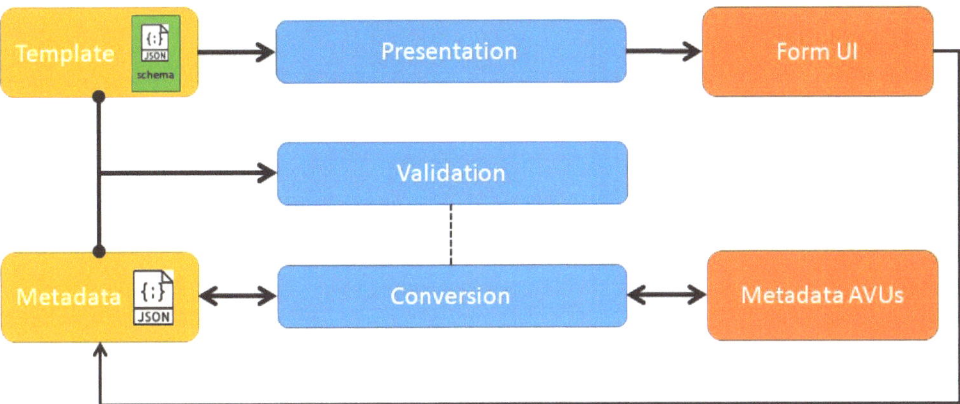

Figure 1. Overview of the schematic layers between JSON, AVUs, JSON-schema and its process in conversion, validation and presentation.

used by Utrecht University, The Netherlands to render metadata entry forms in its data management application Yoda [8]. DataHub at MUMC+ and Maastricht University, The Netherlands seeks to implement a similar solution based on the CEDAR Workbench [9]. DataHub's use case is focused on semantically linked (meta)data. Therefore not only should the JSON structure be governed by a JSON Schema, in addition its vocabulary and structure must conform to the W3C JSON-LD recommendation [10].

Hence our research can be applied on three levels. At the foundation level, a conversion method supports the use of JSON to manage arbitrary compositions of AVUs in iRODS and to exchange these compositions efficiently with client applications (Methods section). The optional second level validates the exchanged JSON against a metadata template defined in JSON Schema. Both the first and second levels are implemented in the iRODS server (Results section). A third level can optionally be implemented as part of a client application. It uses the (same) metadata template for dynamic form-based user interactions. In the Discussion section, the main advantages and disadvantages we found using the proposed methods are discussed. Finally, the research results are summarized and an outlook is provided in the Conclusion section.

METHODS
Bidirectional conversion between JSON and AVU structures

We intend to represent a set of iRODS AVUs as a JSON structure and vice versa and use the serialized data in communications between the iRODS server and client applications.

Design goals

Before creating the conversion method we set five design goals.

1. The conversion method must be a bijective function to ensure lossless conversions between JSON and AVU structures in both directions. Any JSON structure that is compliant with the JSON specification should be supported. The method should provide support for Unicode characters, nested structures, and ordered lists.

2. It must be easy to identify corresponding JSON objects and AVU attributes. This means that, especially for simple JSON structures, it should be trivial to retrieve a JSON element from the AVUs without first back-converting the JSON.

3. The conversion method should be lean and efficient. We seek to avoid an explosion in the number of AVUs as a result of representing a nested JSON structure.

4. The method should be compatible with existing use cases that operate directly on AVUs.

5. The conversion method should be compatible with JSON-LD use cases.

Conversion method specifications

Using the design goals set as requirements we arrived, over several iterations, at a working design for the conversion. An example JSON structure and its converted counterpart in AVUs is listed in Table 1. This example will be used to explain how the conversion method works. Further examples can be found in the online repository[4].

Table 1. Example of a JSON structure and its converted counterpart in AVUs.

```
{
    "title": "Hello World!",
    "parameters": {
        "size" : 42,
        "readOnly" : false
    },
    "authors" : ["Foo", "Bar"],
    "references": [
        {
            "title": "The Rule Engine",
            "doi": "1234.5678"
        }
    ]
}
```

Attribute	Value	Unit
title	Hello World!	root_0_s
parameters	o1	root_0_o1
size	42	root_1_n
readOnly	False	root_1_b
authors	Foo	root_0_s#0
authors	Bar	root_0_s#1
references	o2	root_0_o2#0
title	The Rule Engine	root_2_s
doi	1234.5678	root_2_s

Representation in JSON

Representation in AVUs

The proposed conversion method repurposes the unit field of the AVU to encode JSON variable type and structure information. This also reduces the chance of collisions with existing AVUs that presumably have an empty unit field. Currently, nearly all the iRODS microservices that facilitate AVU operations, for example `msiAssociateKeyValue-PairsToObj`, do not allow rule developers to specify content for the unit field. As a result of this limitation, the unit component of the AVU is hardly ever used at the time of writing.

The AVU unit field comprises of four components, separated by an underscore character, except for the last component where a hash is used. The first component indicates a *namespace* carried by all the AVUs that belong to this set. Conversion operations will only affect AVUs that are part of the selected namespace. In addition, it facilitates that iRODS objects are annotated with multiple JSON structures, each identified by their own namespace. In example Table 1 the namespace is `root`.

The second component is an *object sequence number* that keeps track of AVUs that are part of the same JSON object. The top level JSON object is assigned sequence number "0". In the example this includes `title`, `parameters`, `authors` and `references`. Note that the element `parameters` holds a nested object as its value. The next sequence number "1" is assigned to this nested object and note the sequence number in its (otherwise unused) AVU value component, prefixed by the character o. All elements of the nested object, in this example `size` and `readonly`, have the object sequence number "1" in their unit field.

---

[4] see https://github.com/MaastrichtUniversity/irods_avu_json

An important design goal of the conversion method is the support of different variable types within JSON. AVUs only allow string values, while JSON supports various primitive types. The third component of the AVU unit field is used to indicate the JSON *type* of the value. See table 2 for an overview of the supported types. A special case is the `empty array` type that indicates the presence of an array without any members. To achieve a lean conversion, we only create AVUs to represent *members* of an array. We have to make an exception for an empty array, which otherwise would not have any AVU representation at all. Without this provision, a later conversion from AVU back to JSON would not be able to recreate the empty array.

Type	AVU unit-type	AVU value	Remarks
string	s	The literal string	
object	o + object_id	o + object_id	The AVU value field is not used for conversion
boolean	b	"True" or "False"	
number	n	String value of float or int	
null	z	"."	AVUs do not allow empty values
empty string	e	"."	AVUs do not allow empty values
empty array	a	"."	For convenience during conversion. See text.

Table 2. Overview of JSON variable types and their corresponding type string.

The fourth, optional component of the AVU unit field is used to denote an *ordered index* of the element. This component is separated from its predecessor using a hash character #. The JSON specification includes the array type. This is an ordered list of elements of any type. As AVUs are unordered, the last component of the unit field denotes the array index to maintain order in arrays.

Summarizing, the AVU unit field has been used for the following purposes: 1. defining the JSON namespace, 2. the object sequence number, 3. the value type and 4. the array index. A regular expression for capturing these components of the unit field is shown in Listing 1.

```
^([a-zA-Z0-9_])+_([0-9]+)_([osbanze])((?<=o)[0-9]+)?((?:#[0-9]+?)*)
```

Listing 1. A regular expression to parse the components of the unit field

Validation of JSON structure using a JSON Schema template

The conversion method discussed above allows client applications to store JSON structures efficiently within the iRODS server. This method is agnostic to the structure and the semantics of the metadata that is exchanged. For some use cases this may suffice. Many use cases however require that metadata stored or exchanged is compliant with a certain standard. We will now propose a validation method to fulfill this need.

Design goals

The validation method needs to meet three design goals.

1. It must be able to assess that a *stored or exchanged set* of metadata meets predefined quality levels with respect to structure and semantics as typically documented in metadata standard. The metadata schema can vary per iRODS object. For instance, objects that belong to a data set related to the Geosciences may require geospatial location annotations whereas objects related to History disciplines may depend on chronological classification metadata.

2. It should also be possible to annotate a single object with multiple disjunct sets of metadata. The metadata template can vary per set of metadata.

3. In line with the conversion method, the validation method should again be compatible with existing use cases that operate directly on individual AVUs.

Validation method specifications

For the purpose of validation, we shall consider sets of metadata that annotate an iRODS object rather than individual AVUs. These sets are easily identified by their namespace identifier which is incorporated in the unit component of the AVU. This makes the validation compatible with existing use cases as AVUs without a namespace will not be affected by the validation and protection policies.

We select the draft internet standard JSON Schema to specify a metadata template used to validate sets of metadata [7]. JSON Schema conveniently supports the description of quality properties of individual metadata elements as well as qualities that span across elements, for instance dependency relationships. Incoming metadata will be in JSON representation and can be validated directly against a template. An example of the resulting schema is shown in Listing 2.

```
{ "$id": "http://example.com/myschema.json",
  "$schema": "http://json-schema.org/schema#",
  "type": "object",
  "additionalProperties": false,
  "properties": {
    "title": { "type": "string" },
    "parameters": {
      "type": "object",
      "additionalProperties": false,
      "properties": {
        "size": { "type": "number" },
        "readOnly": { "type": "boolean" }
    }},
    "authors": {
      "type": "array",
      "items": { "type": "string" }
    },
    "references": {
      "type": "array",
      "items": {
        "type": "object",
        "additionalProperties": false,
        "properties": {
          "title": { "type": "string" },
          "doi": { "type": "string" }
}}}}}
```

Listing 2. JSON schema that corresponds to the JSON structure of Table 1.

RESULTS

Conversion method implementation

The implementation has been developed for iRODS version 4.2.x [11]. Since iRODS 4.2 does not expose any microservices to modify the unit field of an AVU triple, custom iRODS microservices have been developed. The conversion scheme has been developed as Python 2.7 and Python 3 module[5] named `irods_avu_json`. The outcome of the conversion of the example JSON is shown in Table 1.

The `irods_avu_json` module has in itself no interaction with or dependency on iRODS. To expose this functionality within an iRODS installation we developed an iRODS ruleset [6]. The developed ruleset uses the Python Rule Engine recently released with iRODS 4.2 to import the irods_avu_json module.

An overview of the functionality being exposed by the ruleset is summarized in Table 3. All ruleset functions allow specifying any iRODS object type (collection, object, user, group or resource) in a similar fashion as that iRODS AVUs can be attached to any iRODS object. Furthermore, all functions also expect the JSON namespace to know which AVU set to operate on.

Validation method implementation

The special *$schema* AVU denotes whenever a JSON Schema is attached to an iRODS object. We implemented two ways for the JSON Schema to be specified in the value field of this AVU. (1) An (public or private) URI pointing to the stored JSON schema. Optional caching of this URI has been implemented for performance reasons. (2) By specifying 'i:' in front of a path the JSON Schema is directly retrieved from within iRODS. Care must be taken that this iRODS object is accessible for anyone allowed to modify the iRODS object to which the JSON Schema is attached. Other methods for storing the JSON Schema could be devised and implemented at a later point.

Note that both the iRODS server and client applications can benefit from using the metadata template to check the validity of any metadata that is being exchanged. Therefore a best practice is that the template reference is an absolute URI and the metadata template itself is available at an internet-accessible location.

Validation of the JSON structure set by `setJsonToObj()` is triggered by the presence of the `$schema` attribute and the same JSON namespace being present on the iRODS object. After retrieving the JSON Schema contents, the validation is performed using the `jsonschema` Python module. Any validation errors will be passed back to the caller of `setJsonToObj()`. To ensure only the full and validated JSON object is being stored first all existing AVUs of the same JSON namespace are removed before the new one are set. This also ensures that any no longer existing parts of the JSON object are removed during a `setJsonToObj()` operation.

The irods_avu_json-ruleset further implements validation of a JSON object by implementing a policy enforcement points (PEPs), which are executed during the modification of AVUs. Whenever a *$schema* AVU is present on the iRODS object and the AVU being modified is part of the same JSON namespace the operation is disallowed. Modification of these AVUs can only be performed through `setJsonToObj()`. Because `setJsonToObj()` would also trigger the same PEPs, the PEPs check whether execution is coming from `setJsonToObj()` and has been validated against the JSON Schema. This is achieved through setting a Python global variable that is preserved in the rule engine memory.

Function	Description
setJsonToObj(object, objectType, jsonNs, json)	Set a JSON to an iRODS object.
getJsonFromObj(object, objectType, jsonNs)	Retrieve a JSON from an iRODS object.
getJsonSchemaFromObj(object, objectType, jsonSchema, jsonNs)	Attach a JSON Schema to an iRODS objec.
setJsonSchemaToObj(object, objectType, jsonNs)	Retrieve a JSON Schema from an iRODS object.

Table 3. Functionality exposed by the irods_avu_json-ruleset

---

[5]see https://github.com/MaastrichtUniversity/irods_avu_json
[6]see https://github.com/MaastrichtUniversity/irods_avu_json-ruleset

DISCUSSION

We successfully used the conversion method and the accompanying validation method in several pilot use cases. We found several limitations and problems with the current implementation of the conversion and validation method.

The implementation of the conversion method currently requires all existing AVUs relating to a JSON namespace to be removed before the new JSON is added. This may lead to performance issues with very large JSON structures. Furthermore, the addition of all JSON related AVUs is not an atomic operation, meaning that collisions may occur if multiple clients modify the same iRODS object at once.

The current implementation of validation uses the metadata PEPs to prevent any non-validated AVUs to be created or modified. These PEPs directly wrap around their AVU modification microservice counterparts. Therefore a single microservice call can, using a wildcard, operate on multiple AVUs. This means logic created for the PEPs is rather convoluted. Furthermore, the chosen implementation of using a global Python variable to bypass the PEPs when `setJsonToObj()` is being called breaks when the call for `setJsonToObj()` is initiated from a catalog consumer instead of the catalog provider. This is because in that case of the call being initiated from the catalog consumer the global variable is not in memory when the PEP is being executed on the catalog provider.

The performance issue, the non-atomic operation of the current conversion and the issue with the PEPs can all be tackled by the introduction of a multi-AVU atomic core iRODS functionality. Such a microservice could handle the entire operation of converting a JSON structure and its validation at once.

While not directly shown in this work the use of JSON Schema can be extended beyond the validation level that is currently implemented. As mentioned before, an important feature is the auto-generation of web forms from the JSON Schema. Several implementations of libraries capable of this functionality exists and we have explored several of those. Furthermore, we envision that the use of JSON Schema for presentation can be further extended to for example auto-generated search forms.

CONCLUSION

We set out to add a generic toolset to iRODS for handling richer, templated and validated metadata. We have developed a bidirectional conversion between JSON and AVUs and validation provided through JSON Schema. Both methods have been validated to meet their design goals via a proof of concept followed by an application-level implementation. The methods developed provide new starting points for iRODS developers in facilitating research data management that adheres to the FAIR principles.

ACKNOWLEDGMENTS

This work was sparked by the discussions in the iRODS metadata template working group and we would like to thank them for their feedback. We would like to thank all members of the DataHub Maastricht team for their helpful feedback and comments. Finally we thank Raimond Ravelli, Peter Peters and Michel Dumontier for their critical reading of this manuscript.

REFERENCES

[1] M. R. Munafò, B. A. Nosek, D. V. Bishop, K. S. Button, C. D. Chambers, N. P. Du Sert, U. Simonsohn, E.-J. Wagenmakers, J. J. Ware, and J. P. Ioannidis, "A manifesto for reproducible science," *Nature human behaviour*, vol. 1, no. 1, p. 0021, 2017.

[2] C. L. Borgman, *Big Data, Little Data, No Data, Christine L. Borgman, The MIT Press, Cambridge, MA (2015), Xxiv, 383 p. $70.00, ISBN: 978-0-262-02856-1*. Elsevier, 2015.

[3] M. D. Wilkinson, M. Dumontier, I. J. Aalbersberg, G. Appleton, M. Axton, A. Baak, N. Blomberg, J.-W. Boiten, L. B. da Silva Santos, P. E. Bourne, and others, "The FAIR Guiding Principles for scientific data management and stewardship," *Scientific data*, vol. 3, 2016.

[4] R. Moore, "Towards a theory of digital preservation," *International Journal of Digital Curation*, vol. 3, no. 1, 2008.

[5] ECMA, "ECMA-404: The JSON Data Interchange Syntax," *Ecma International*, vol. Standard, no. Second edition, 2017.

[6] T. Bray, "IETF RFC 8259: The JavaScript Object Notation (JSON) Data Interchange Format," *Internet Engineering Task Force (IETF)*, 2017.

[7] H. Andrews and A. Wright, "JSON Schema A Media Type for Describing JSON Documents," *Internet Engineering Task Force (IETF)*, vol. Internet-Draft, Mar. 2018.

[8] T. Smeele and L. Westerhof, "Using iRODS to manage, share and publish research data: Yoda," in *Proceedings of the 2018 iRODS User Group Meeting*, (Durham NC), University of North Carolina, June 2018.

[9] R. S. Gonçalves, M. J. O'Connor, M. Martínez-Romero, A. L. Egyedi, D. Willrett, J. Graybeal, and M. A. Musen, "The CEDAR Workbench: an ontology-assisted environment for authoring metadata that describe scientific experiments," in *International Semantic Web Conference*, pp. 103–110, Springer, 2017.

[10] M. Sporny, D. Longley, G. Kellogg, M. Lanthaler, and N. Lindström, "JSON-LD 1.0," *W3C Recommendation*, vol. 16, p. 41, 2014.

[11] A. Rajasekar, R. Moore, M. Wan, and W. Schroeder, "Policy-based Distributed Data Management Systems," *Journal of Digital Information*, vol. 11, no. 1, 2010.

iRODS at KTH and SNIC – Status and Prospects

Ilari Korhonen
KTH Royal Institute of Technology
Stockholm, Sweden
ilari@kth.se

ABSTRACT

The current state of iRODS operations at KTH PDC Center for High Performance Computing is presented with a two-pronged approach. Firstly, the status of our national iRODS-based data infrastructure is laid out with respect to both the deployment at KTH PDC as well as our collaboration with our partner center NSC at Linköping University. We are currently a distributed operation between two centers, with the possibility of more Swedish HPC centers to join in the future. Secondly, our development efforts for an HPC-adjacent iRODS deployment at our center is discussed. The focus here is to provide access to several tiers of iRODS storage environments in the high performance computing environments available at our center, in a secure, efficient, low-latency and high-throughput configuration. In addition to providing iRODS clients in our compute cluster to enable access to (federated) iRODS grids while using (federated) Kerberos authentication between the administrative domains, we are also collaborating with the iRODS Consortium on the testing of a new iRODS Lustre interface. Eventually this would enable us to provide Lustre-backed iRODS resources for storage tiering within the local iRODS grid, for the staging of HPC data into and especially out of the compute filesystem. This combined with federated access to several tiers of storage resources and the ability to publish data out of iRODS, possibly in the future referenced by persistent identifiers, gives a working solution for research data lifecycle management.

An authentication solution for iRODS based on the OpenID Connect protocol

Claudio Cacciari
CINECA - Interuniversity
Consortium
Casalecchio di Reno
(BO) - Italy
c.cacciari@cineca.it

Giuseppa Muscianisi
CINECA - Interuniversity
Consortium
Casalecchio di Reno
(BO) - Italy
g.muscianisi@cineca.it

Michele Carpené
CINECA - Interuniversity
Consortium
Casalecchio di Reno
(BO) - Italy
m.carpen@cineca.it

Mattia D'Antonio
CINECA - Interuniversity
Consortium
Casalecchio di Reno
(BO) - Italy
m.dantonio@cineca.it

Giuseppe Fiameni
CINECA - Interuniversity
Consortium
Casalecchio di Reno
(BO) - Italy
g.fiameni@cineca.it

ABSTRACT

We are going to describe an authentication solution for iRODS based on the OpenID Connect (OIDC) protocol. In the context of European data infrastructures, like EUDAT, and projects, like EOSC-hub, iRODS must interoperate with other services, which support OIDC and OAuth2 protocols. The typical usage workflows encompass both direct user interaction via icommand and other clients and service-to-service interaction on behalf of the user. While in the first case we can rely on the already existing iRODS OpenID plugin, in the second one it is not possible because of two main reasons. The first is that the service-to-service process implies that the user is not requested to generate a token for iRODS, but that iRODS is able to re-use an existing token from another service. The second is that the other service needs to get access to iRODS using multiple authentication protocols in a dynamic way, not fixing one of them in the configuration. For example we have instances of DavRODS that allow to log-in via plain username and password or via OIDC token. In order to achieve those results we implemented a Pluggable Authentication Module (PAM), which allows iRODS to accept an OIDC token, re-using the password parameter of the PAM based authentication, validate it against an Authentication Service and map the user to a local account relying on the attributes provided back by the Authentication Service, once validated the token. Given the flexibility of the PAM approach, in this way it is possible to stack multiple PAM modules together, enabling a single iRODS instance to support multiple OIDC providers and even to create dynamically the local accounts, without any pre-configured mapping.

Keywords

OpenID Connect, B2ACCESS, B2SAFE, PAM, iRODS, authentication.

INTRODUCTION

IRODS was adopted years ago in the EUDAT CDI, Collaborative Data Infrastructure [1], as main component of the service for the long term data preservation and the policy driven data management, called B2SAFE [2]. Within the EUDAT CDI ecosystem the authentication is managed through an OpenID Connect [3] Server, called B2ACCESS. In order to support such protocol, we have developed a Pluggable Authentication Module (PAM) that enables iRODS to authenticate a user with an OIDC token and to map it to an iRODS's local account. The implemented mechanism relies on the PAM support of iRODS and assumes that the user has already obtained a valid token before logging in to iRODS (Figure 1). While no assumption is made about the corresponding local (to the iRODS system) account. It

iRODS UGM 2019 June 25-28, 2019, Utrecht, Netherlands

can be created in advance by the administrator and explicitly mapped to the global user account (the one associated to the OIDC token). Or it can be dynamically created at the first login attempt through a shell script triggered at the end of the PAM modules' stack.

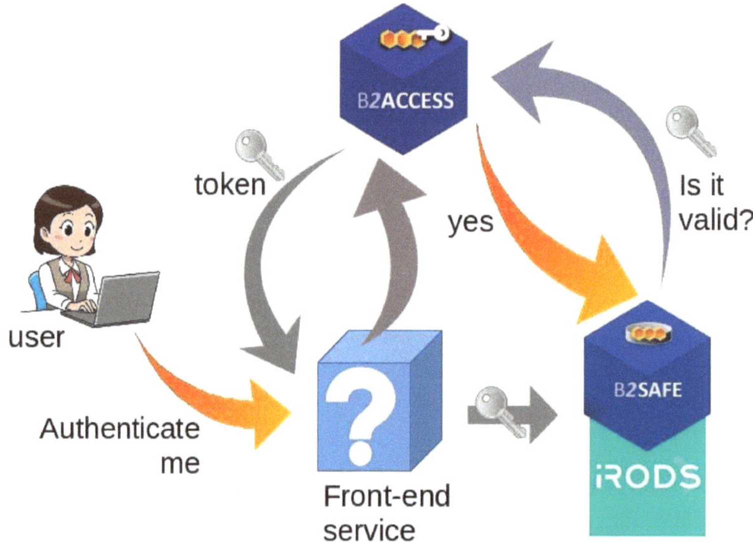

Figure 1. Overview of the main authentication scenario.

B2SAFE is implemented as a set of iRODS rules, python scripts and microservices and it is distributed as an optional package to be deployed on top of an existing iRODS instance.

In the next chapters we will explain which is the problem we have addressed, the details of the solution we have implemented, its benefits and limits and the perspective.

THE AUTHENTICATION ISSUE

The authentication flow

Our system, the EUDAT CDI, is composed by multiple distributed services, most of which belonging to different administrative domains. Each service works standalone, offering interfaces as entry points for the users, and, at the same time, can be the back-end of other services of the same system. Naturally we want to offer a user-friendly environment, therefore we adopted a federated identity approach, as defined by Gaedke, Johannes and Nussbaumer [4] and Chadwick [5], to support a single identity for the user across the whole infrastructure. There is a central authentication service (B2ACCESS), which acts both as OpenID Provider (OP) and as bridge for multiple Identity Providers (IdP). The typical authentication flow to get access to a service is listed below:

1. The user login to the wanted service.

2. The service redirect the user to the OP via OIDC protocol.

3. The user authenticates himself against the OP using the credentials and the identity associated to one of his IdPs.

4. The service receives a response from the OP and based on that, it allows the user to perform the required action and/or get access to the desired resource.

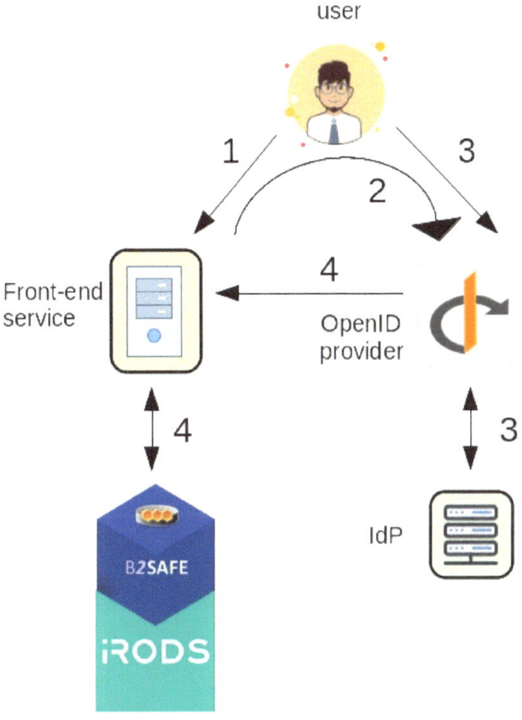

Figure 2. Authentication flow to get access to B2SAFE service.

In Figure 2 is shown the authentication flow to get access to the B2SAFE service.

We are assuming here to rely on the OIDC authorization code flow [6].

The OIDC authorization code flow

We will now describe more in details the authorization code flow [7] ("OpenID Connect explained | Connect2id", n.d.):

1. The service initiates user authentication by redirecting the browser to the OAuth 2.0 authorization endpoint of OP.

2. At the OP, the user will typically be authenticated prompting the user to login. After that the user will be asked whether they agree to sign into the service.

3. The OP will then call the service's callback URI with an authorization code (on success) or an error code (if access was denied, or some other error occurred, such a malformed request was detected).

4. The service must use the code to proceed to the next step, exchanging the code for the ID token. The authorization code is an intermediate credential, which encodes the authorization. It is therefore opaque to the service and only has meaning to the OP server. To retrieve the ID token the service must submit the code to the OP. The code-for-token exchange happens at the token endpoint of the OP.

5. The service has been previously registered with the OP and has got a client ID and a secret. Both are passed via the Authorization header in the request to obtain the ID token. On success the OP will return a JSON object with the ID token, an access token and an optional refresh token.

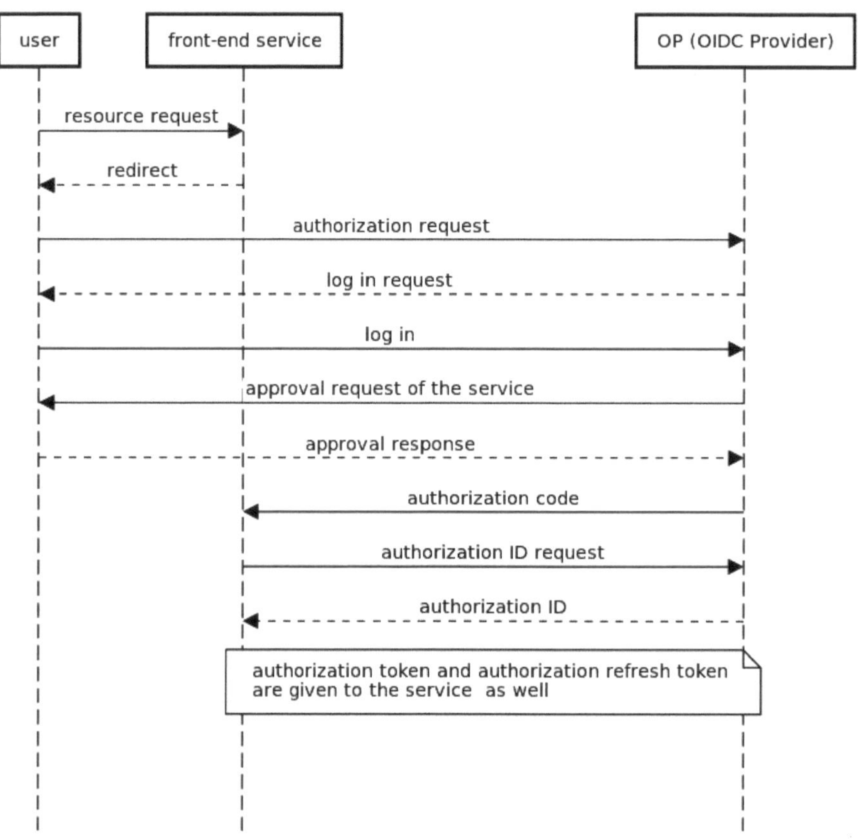

Figure 3. Authorization code flow sequence diagram.

The issue

The aforementioned flow is supported by the services of our system. IRODS itself can be configured to support it, using the OpenID plugin [8]. However when we want to implement a workflow chaining together a first service, that we will call front-end, and B2SAFE as back-end, we have troubles. Usually the user wants to have access to B2SAFE, and hence to iRODS, because she wants to download data from or upload data to it. We assume that the user logins into the front-end service using the OIDC protocol and her federated identity. At this point we need that the user gets access to the B2SAFE service with the same identity, which has to be mapped to an iRODS local account. How the front-end service can pass the user identity credentials to B2SAFE so that it can authenticate the user? We could ask the user to login explicitly to B2SAFE, but this is exactly what we want to avoid. It would be possible to use a third software component, like a Web portal or a workflow manager, which would impersonate the user in front of the services, but they are not available in our system. In fact the services are connected directly to each other through their APIs. In the next chapter we will describe the proposed solution.

THE PROPOSED SOLUTION

After that the user logs in successfully into the front-end service, this last one receives the ID token, the access token and the refresh token. Therefore our proposal is to rely on those tokens to authenticate the user against B2SAFE and in particular on the access token. In order to authenticate the user, B2SAFE relies on iRODS's authentication mechanisms. But they do not support OIDC tokens as authentication credentials, thus we need to extend them. In particular we need to satisfy two requirements:

1. The user identity credentials must be validated

2. The user federated identity must be mapped to an iRODS's local account.

The first point can only be achieved presenting the access token to the OIDC provider, B2ACCESS in our case, which has issued the token. OpenID Connect specifies a set of standard claims, or user attributes. They are intended to supply the client with consented user details such as email, name and picture, upon request. The OIDC provider's UserInfo endpoint returns them. Submitting a request containing the access token to the UserInfo endpoint, it is possible to validate the token and to get some user attributes. In our case, B2ACCESS can provide us the email of the user. Assuming the email address is unique for each user, we can use it to map the federated identity of the user to an iRODS local account and satisfy in this way the second requirement.

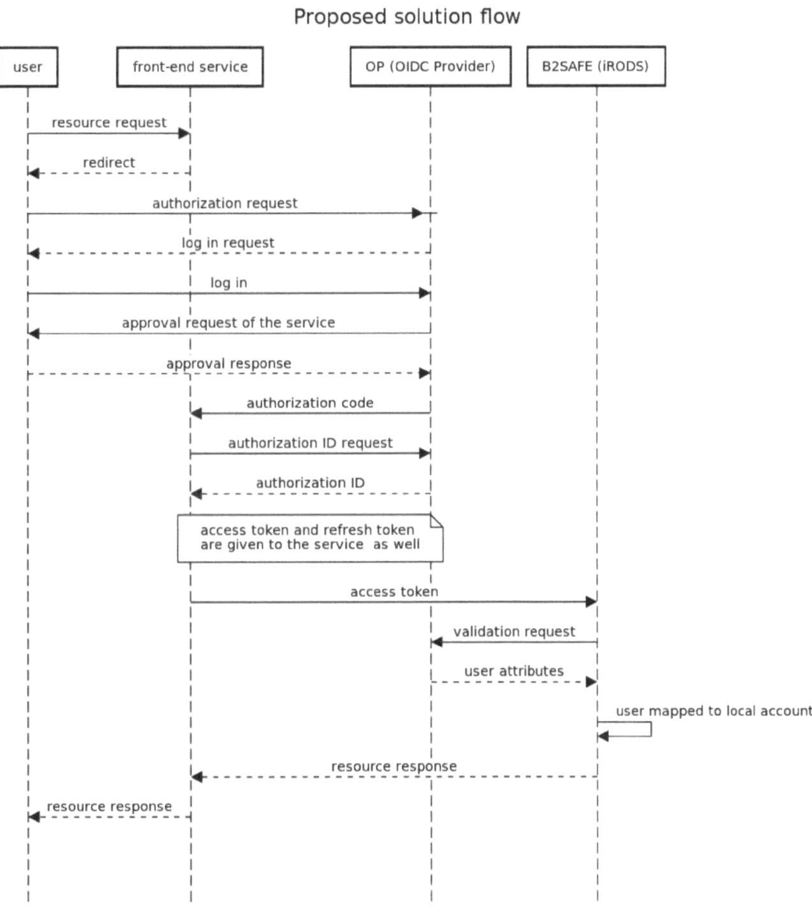

Figure 4. New proposed authorization code flow sequence diagram.

Implementation

We have implemented the proposed solution as a PAM module [9], written in C, which needs to be compiled and configured. The choice of PAM as framework permits to have more flexibility compared to other solutions, like implementing a new iRODS authentication plugin, as we will show in the next paragraphs. To enable the new module, the iRODS PAM configuration file (/etc/pam.d/irods) has to include the following line:

```
auth sufficient pam_oauth2.so <path to configuration file> key1=value2 key2=value2
```

Where pam_oauth2.so is the name of the compiled new PAM module and key1 and key2 optional parameters. Assuming the path to the configuration file is /etc/irods/pam.conf, its content looks like this:

```
#OAuth2 url for token validation
token_validation_ep = "https://b2access.eudat.eu/oauth2/userinfo"
#OIDC attribute key to identify the attribute in the response used to match the login username
login_field = "email"
# path to user map file
user_map_file = "/etc/irods/user_map.json"
```

Where:
token_validation_ep: it is the OIDC UserInfo endpoint.
login_field: it is the attribute that needs to be checked for the user mapping.
user_map_file: it is the path to the file defining the mapping rules.
The user_map_file is formatted as a json document and looks like this:

```
{
  "roberto": ["roberto@email.it", "r.mucci@email.it"],
  "claudio": ["c.cacc@email.com", "claudio@email.it", "c.cacciari@email.it"],
  "paolo": ["paolo@email.com"]
}
```

A list of email addresses, on the right, can be mapped to a single iRODS account, on the left.

How it works

Since we have decided to rely on the iRODS PAM mechanism, the user, or better, in our case, the front-end service has to login with the PAM authentication method, but instead of the password of the iRODS local account, it uses the access token. The PAM module pam_oauth2.so receives the token and issues a request to the B2ACCESS's token_validation_ep. If the token is valid, the response looks like this:

```
{
  "email":"roberto@email.com",
  "token_type":"Bearer",
  "exp":1520001942,
  "iat":1519998342,
  [ ... ]
}
```

And the front-end service will get access as the local user "roberto". Enabling the debug log of the PAM framework, the following lines will appear in the syslog:

```
Token validation EP: https://b2access.eudat.eu/oauth2/userinfo
Jun  1 09:26:34 localhost irodsPamAuthCheck: username_attribute: email
Jun  1 09:26:34 localhost irodsPamAuthCheck: user_map_path: /etc/irods/user_map.json
Jun  1 09:26:34 localhost irodsPamAuthCheck: Searching for user: roberto
Jun  1 09:26:34 localhost irodsPamAuthCheck: Found local mapping item: roberto@email.it
Jun  1 09:26:34 localhost irodsPamAuthCheck: Found local mapping item: r.mucci@email.it
Jun  1 09:26:34 localhost irodsPamAuthCheck: pam_oauth2: Found user mapping array for user
                                             roberto
Jun  1 09:26:34 localhost irodsPamAuthCheck: pam_oauth2: user mapping item returned by B2ACCESS:
                                             roberto@email.com
Jun  1 09:26:34 localhost irodsPamAuthCheck: pam_oauth2: successfully mapped item to local
                                             iRODS user
Jun  1 09:26:34 localhost irodsPamAuthCheck: pam_oauth2: successfully authenticated by B2ACCESS
```

BENEFITS AND LIMITS

The proposed solution solves the aforementioned authentication issue because it allows to reuse the OIDC tokens without requiring the user to login again when the B2SAFE service is involved as back-end in a workflow. The front-end service can access B2SAFE using the same federated identity of the user and iRODS is able to validate the OIDC access token associated to that identity and to map it to a local account.

Another benefit is the flexibility in the account mapping. Thanks to the pam_exec, a PAM module that can be used to run an external command, it is possible, after a successful authentication, to execute a script which calls the iRODS command to create a user on the fly. Therefore it is possible to create users with the same username of the federated identity or just implement a pool of accounts which are allocated dynamically and so on.

Moreover this approach allows to support multiple OIDC providers at the same time, stacking multiple modules pam_oauth2.so in the same file /etc/pam.d/irods, each one pointing to a different configuration file. Indeed this is true for multiple PAM modules in general. For example it is possible to stack together pam_oauth2.so and the LDAP module, so that if one authentication method fails, the credentials are passed to the next one.

One limit of the described solution is the need to know the local iRODS username at the login time because this is required by the iRODS PAM authentication mechanism. This reduces the flexibility of the user mapping procedure. It is still possible to create dynamically a user account, but the username must be predictable.

More important is the fact that the access token has a limited lifetime, which is, in our case, set to one hour. The OIDC protocol includes a refresh token to allow the client, the front-end service in our case, to extend the lifetime when needed. However iRODS cannot do it because the lifetime can be extended only by the same client that has requested it the first time. Therefore the front-end service must support the refreshing of the tokens, possibly in a transparent way for the user.

USE CASES

The solution has been tested with two front-end services in the context of the EOSC-hub project [10]. The first one is an HTTP interface [11], which exposes some functions to upload/download data using the iRODS python library [12]. The HTTP API serves users that have a federated identity and users that have just a local identity, from the same instance. This is possible thanks to the flexibility of the PAM module.

The second one is a data management tool called DataHub [13], which is able to mount external storage if this storage is exposed through a WebDAV interface. In this case the workflow is composed by three services because DataHub is connected to the WebDAV interface [14], which is deployed in front of iRODS. DataHub takes care of the initial user authentication and hence of the token refreshing.

CONCLUSION

We have described the implementation of a solution to add the support of the OpenID Connect protocol to iRODS relying on its PAM authentication mechanism. The solution is flexible enough to enable different use cases and satisfies the requirements of the single identity of the user and of the validation of the OIDC credentials. Future developments can be envisioned about the user mapping mechanism. Currently it is just based on a static map represented as a json document. However that approach does not scale very well. It would be better to map the user in dynamic way based on the attributes. For example supporting the use of regular expressions to define matching rules, we could map all the users belonging to a certain organization to a single iRODS local user or group. In this way, even if new members join the organization, it would not be necessary to add them manually to the user map, like we do now.

ACKNOWLEDGMENTS

The implementation of the PAM module for the OIDC protocol started forking the code of the OAuth2 PAM module by Alexander Kukushkin [15].

REFERENCES

[1] EUDAT CDI, Collaborative Data Infrastructure, `https://www.eudat.eu`

[2] B2SAFE service, `https://www.eudat.eu/b2safe`

[3] OpenID connect, `https://openid.net/connect`

[4] Gaedke, M., Meinecke, J., Nussbaumer, M.: A modeling approach to federated identity and access management. In: Special interest tracks and posters of the 14th international conference on World Wide Web - WWW '05. ACM Press. `https://doi.org/10.1145/1062745.1062916` (2005)

[5] Chadwick, D. W.: Federated Identity Management. In: Foundations of Security Analysis and Design V, pp. 96-120. Springer Berlin Heidelberg. `https://doi.org/10.1007/978-3-642-03829-7_3` (2009)

[6] OpenID authorization code flow, `https://openid.net/specs/openid-connect-core-1_0.html`

[7] OpenID connect, `https://connect2id.com/learn/openid-connect`

[8] OpenID iRODS plugin, `https://github.com/irods-contrib/irods_auth_plugin_openid`

[9] PAM module developed, `https://github.com/EUDAT-B2SAFE/pam-oauth2`

[10] EOSC-hub project, `https://www.eosc-hub.eu`

[11] HTTP interface, `https://github.com/EUDAT-B2STAGE/http-api`

[12] iRODS python library, `https://github.com/irods/python-irodsclient`

[13] DataHub, `https://www.eosc-hub.eu/services/EGI%20DataHub`

[14] DavRODS - WebDAV interface, `https://github.com/UtrechtUniversity/davrods`

[15] OAuth2 PAM module, `https://github.com/CyberDemOn/pam-oauth2`

Asynchronous file handling with iRODS tape resources

Arthur Newton
SURFsara
Amsterdam, Netherlands
arthur.newton@surfsara.nl

ABSTRACT

Tiered storage systems comprised of a disk cache staging area and a tape library are cost-effective solutions ideal for long-term storage.

Last year, we presented a way to build a tiered storage system which employs tape storage in the backend transparent to iRODS. Since our tape archive system already has a disk cache, we explicitly did not make use of an iRODS compound resource for integration which would have required an additional cache layer.

Tape storage is inherently asynchronous, meaning that data can reside in different states, online on disk cache, or offline only on tape. If the data is offline, it is not readily available and needs to be staged to disk. Our solution in iRODS can (automatically) trigger state transitions between offline and online. However, the user still experienced the asynchronicity of the different states of data, which left room for improvement in the user friendliness of handling such data.

To fill this gap, we implemented a set of command line applications which makes it easier for the user to download, upload, and retrieve information about the state of data. The iRODS python client provides the base of the tool, which alleviates the need for icommands or the DMF tape tools and as such also broadens the compatibility on different systems. The application is split into a set of CLI tools and a daemon-like application that handles requests and file transfers in the background. The daemon is automatically spawned as a non-root process upon the first request and stopped when idle for a specific time.

The command line tool can be extended to other types of storage resources with similar asynchronous staging of data. Additionally, the performance can possibly be improved by allowing for parallel transfer of data.

iRODS UGM 2019, June 25-28, 2019, Utrecht, Netherlands

A GA4GH Data Repository Service for native iRODS

Mike Conway
NIH / NIEHS
Durham, NC, USA
mike.conway@nih.gov

ABSTRACT

This paper and presentation will premier implementation of a GA4GH Data Repository Service, (formerly the GA4GH Data Object Service), that can run on a base iRODS server. The Data Repository Service implementation uses standard iRODS collections to house a Data Bundle, and utilizes Attribute-Value-Unit metadata to mark data objects and bundles with auxiliary information.

Using this service allows iRODS to integrate with workflow and data access services in the area of genomics and bio-sciences that follow the emerging standards of the GA4GH consortium. These standards have been identified in the NIH Commons effort as an important interoperability standard.

The development approach of this implementation sets a very low barrier of entry and allows any genomic data set stored in iRODS to be exposed via the GA4GH Data Repository Service API. A presentation, paper, and source code release are planned for the User Group Meeting.

iRODS UGM 2019, June 25-28, 2019, Utrecht, Netherlands

SODAR – the iRODS-powered System for Omics Data Access and Retrieval

Mikko Nieminen
Berlin Institute of Health
Berlin, Germany
mikko.nieminen@bihealth.de

ABSTRACT

In the past years, a growing number of high-throughput omics assays in the areas of genomics, proteomics, and metabolomics have become widespread in life science research. This creates increasing demand for handling the large amounts of data as well as models for the complex experimental designs. Further challenges include the FAIR principles for making the data findable, accessible, interoperable, and reusable. Collaboration between multiple institutes further complicates data management.

Here we present SODAR (System for Omics Data Access and Retrieval), our effort of fulfilling these requirements. The modular system allows for the curation of complex studies with the required meta data as well as for the storage of large bulk data. To facilitate effective and efficient data management workflows, SODAR provides project-based access control, a web-based graphical user interface for data management, programmatic data access, ID management for study objects as well as various tools for omics data management.

The system is based on open source solutions. iRODS is used for large data storage while Davrods allows for providing access through the widely supported HTTP(S) protocol (e.g., for integration with the IGV software). Graphical interfaces and APIs are implemented in Python using the Django framework. Our data model is based on the ISA-Tools data model ISA-tab is used as the meta data file exchange format. A transaction sub system integrates activities spanning both data and meta data. Core parts of SODAR are available as reusable libraries for creating project-based data management systems that share access control with SODAR.

We will demonstrate our flag ship rare disease genetics use case, starting from bulk data import to the browsing of study design and metadata and interacting with the data through IGV.

A beta version of SODAR is currently deployed in our institutes. The system will be made available as open source under a permissive license.

iRODS UGM 2019, June 25-28, 2019, Utrecht, Netherlands

iRODS in context: Exploring integrations between iRODS and OwnCloud

Hylke Koers
SURFsara
Amsterdam, Netherlands
hylke.koers@surfsara.nl

ABSTRACT

Within the Netherlands, iRODS is gaining substantial traction with universities and other research institutes as a tool to help manage large amounts of heterogeneous research data. In this context, iRODS is usually used as middleware, providing value through data virtualization, metadata management and/or rule-driven workflows. This is then typically combined with other tools and technology to fully support the diverse needs of researchers, data stewards, IT managers, etc.

While integrations with other RDM tools are facilitated by iRODS' flexibility, a significant amount of work is usually still required to develop and test them with users in their specific context. For this reason, SURF – as the collaborative ICT organisation for Dutch education and research – sees a role for itself to spearhead the development of such integrations as that effectively means pooling of resources which lowers the collective development cost and accelerates the pace of adoption.

In this contribution, we will focus on a recent project undertaken by SURF to explore the integration between OwnCloud and iRODS. OwnCloud is an open-source, "sync and share" solution to manage data as an individual or as a research team. OwnCloud is the technology behind two successful existing SURF products: SURFdrive and Research Drive. Offering a GUI, versioning, off-line sync and link-based sharing, its functionality is in many ways complementary to iRODS. This makes integrating the two technologies attractive, yet there are several challenges in terms of file inventory synchronization, metadata management, and access control. In the demo, we'd like to share how we have addressed these challenges and discuss a proposed way forward.

As an outlook into future work, this integration could be extended to support seamless publication of research data in trusted, long-term data repositories. Existing data publication workflows have many common tasks, but also significant variance in the "details" of how these tasks are stringed together and how they need to be operationalized. To address this balance, we are exploring an approach that essentially abstracts data publication tasks into an overarching workflow framework, so as to allow for flexibility yet also benefit from standards and common patterns.

iRODS UGM 2019, June 25-28, 2019, Utrecht, Netherlands

iRODS and use case of Bristol-Myers Squibb to manage genomics data

Oleg Moiseyenko
Bristol-Myers Squibb Company
Princeton, NJ
Oleg.Moiseyenko@bms.com

ABSTRACT

This presentation will discuss how iRODS helps Bristol-Myers Squibb manage petabytes of NGS data, synchronizing it from different on-premise locations with AWS Cloud store and challenges it presents.

In the days of high speed internet and cloud computing, the old paradigms in drugs discovery went through significant changes. The shifts in IT landscape and medicine economics force "big pharma" companies to seek better routes for innovation and efficiencies. Since DNA has gone digital, various scientific communities around the globe routinely run multiple tests, take high-resolution medical images, and use big data in health research on daily basis.

New cloud computing infrastructure contributes to swift increases in research partnerships in bioanalysis via collaboration consortia, which at the end leads to data fragmentation in terms of sources, data types, and storage. In addition, this data should be also securely stored, well maintained through the lifetime, and access-controlled in accordance with latest local regulations and data compliance requirements.

iRODS UGM 2019, June 25-28, 2019, Utrecht, Netherlands

iRODS Capabilities: Indexing and Publishing

Jason Coposky
Renaissance Computing Institute (RENCI)
UNC-Chapel Hill
jasonc@renci.org

ABSTRACT

This presentation shares the fourth and fifth iRODS Capabilities with working code. Indexing and Publishing are very similar in that they instrument the iRODS Zone to watch for particular metadata and then engage external API endpoints to populate external systems. The candidate code demonstrates indexing via Elasticsearch and publishing via Data.World.

NFSRODS: Presenting iRODS as NFSv4.1

Kory Draughn
Renaissance Computing
Institute (RENCI)
UNC Chapel Hill
korydraughn@renci.org

Terrell Russell
Renaissance Computing
Institute (RENCI)
UNC Chapel Hill
unc@terrellrussell.com

Alek Mieczkowski
Renaissance Computing
Institute (RENCI)
UNC Chapel Hill
info@irods.org

Jason Coposky
Renaissance Computing
Institute (RENCI)
UNC Chapel Hill
jasonc@renci.org

Mike Conway
NIH / NIEHS
mike.conway@nih.gov

ABSTRACT

NFSRODS[1] is a new iRODS[2] client that presents the iRODS logical namespace as NFSv4.1[3]. This allows administrators to expose the iRODS namespace, for both reads and writes, to any software or users who do not know how to speak the iRODS protocol. Existing scripts, tools, or hardware can traverse the mountpoint as expected, but while doing so, they also invoke all the server-side policy that iRODS provides and enforces. The v0.8.0 release has preliminary iRODS multi-owner access mapped to traditional Unix permissions, but will be updated for NFSv4 ACLs before v1.0.

Keywords

iRODS, client, NFS, NFSv4, data management

INTRODUCTION

NFSRODS v0.8.0 is informed by earlier work from Brazil in 2016. NFS-RODS[4] reported that the mapping from NFS calls and attributes to iRODS attributes was the hardest part. In the end, this NFSv3[5] implementation was not expressive enough to present a lossless interface into the iRODS namespace. iRODS needed an NFSv4 server to work with before more work could be done.

As a summer project in 2018, the iRODS Consortium began an investigation and early coding for what would become today's NFSRODS. It was envisioned to be built atop NFS4J[6] and be a pure Java application to be run in a JVM, probably within a Docker[7] container. The summer ended with a prototype that included Kerberos[8] for authentication and handled the protocol translation from NFSv4.1 to iRODS pretty well. Deploying this version was very complicated and required deep knowledge of Kerberos internals and configuration.

It was decided to trust the deployment environment for authentication and remove the Kerberos layer. Deployment became much simpler and early testing with stakeholders in enterprise environments suggested this was the correct path forward.

At the time of publication, the NFSRODS server is released as v0.8.0 and provides a lossy permission mapping for multiple iRODS owners into the standard traditional Unix permission model.

Future work will use NFSv4 ACLs to provide a lossless bidirectional mapping between the multiple ownership and permission models of NFSv4.1 and iRODS. This will be released as NFSRODS version 1.0.

iRODS UGM 2019 June 25-28, 2019, Utrecht, Netherlands

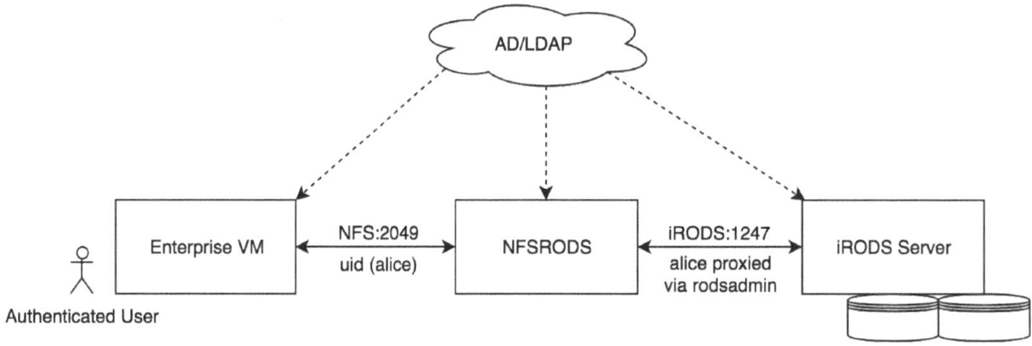

Figure 1. NFSRODS assumes an authenticated user without sudo access within the Enterprise VM.

ARCHITECTURE

The core of NFSRODS has two components. The NFS server side of NFSRODS is provided by NFS4J and is largely lifted directly from the open source project. NFS4J has a plugin architecture for its VirtualFileSystem and allows for other technologies to provide the filesystem interface. The iRODS Jargon client library[9] is used to implement an iRODS VirtualFileSystem. Jargon implements the iRODS protocol and communicates as an iRODS client.

With Kerberos removed, the security model of NFSRODS deployment makes a few assumptions about its environment. First, the usernames and UIDs must be consistent from the mountpoint, to the NFSRODS server, to within the iRODS catalog. The NFS connection between the mountpoint and NFSRODS communites which user is requesting access by unix UID. If `alice` (UID 509) makes a request to `ls` within the mountpoint, the NFSRODS server sees a request from UID 509 only. The NFSRODS server must be able to map the incoming UID to an iRODS username. This is done by the OS and uses the standard `/etc/passwd` and `/etc/shadow` files. Therefore, these must be kept in sync across the different machines in the system. It is assumed that this will be handled via external systems, most usually LDAP. The NFSRODS server maps the incoming request to an iRODS request which uses the matching username. It is assumed that the mechanism keeping the UIDs and usernames consistent is also keeping the list of users within the iRODS catalog consistent.

With this model, it is very important to note that any user with `sudo` rights on the Enterprise VM can become any other user, and therefore gain access to iRODS as that other user. It is recommended that there be no `sudo` rights available on the Enterprise VM where the mountpoint is accessible to the end user.

PERMISSIONS

Traditional Unix permissions[10] are based on rights granted to the three classes of `owner`, `group`, and `world`. Settings for each of these three classes can be `read`, `write`, `execute`, or none. A particular file can have one owner, be part of one group, and then have additional permissions set through the `world` class. This is a very rich and well-understood mechanism for sharing rights to files.

iRODS has a multi-owner model with a linear permission system. Permissions can be set to `OWN`, `WRITE`, `READ`, or none. `OWN` has all the rights of `WRITE` and can delete files and set permissions for others. `WRITE` has all the permissions of `READ` and can write the data or metadata for a data object. `READ` can see the contents of a data object.

Mapping between these two different permissions systems is not straightforward and has proven the most complex part of building NFSRODS.

iRODS Permission	Collection as Directory	Data Object as File
OWN	`drwx-----x`	`-rw-------`
WRITE	`d--x---rwx`	`-------rw-`
READ	`d--x---r-x`	`-------r--`
NULL	`d--x-----x`	`----------`

Table 1. NFSRODS v0.8.0 Mapping of iRODS Permissions to traditional Unix permissions

NFSRODS v0.8.0's mapping of iRODS Permissions to traditional Unix permissions is shown in Table 1. If an iRODS Collection is viewed from within an NFSRODS mountpoint, the directory is shown with the current user as Unix owner if the user is **an** iRODS owner. If this is the case, then the traditional Unix permissions are shown with full permissions `rwx`. If an iRODS data object is viewed from within an NFSRODS mountpoint, then the traditional Unix permissions are shown with permissions `rw`.

Non-owner (aka `WRITE` and `READ`) permissions are mapped, per viewer, as world permissions. Collections are always mapped with the `execute` bit for world, to allow traversal through the namespace.

This mapping works for non-admin and non-sysadmin users of NFSRODS. Access is granted correctly and disallowed correctly across the various combinations of multi-ownership.

However, this presentation of iRODS permissions through the use of `world` permissions proved too alarming and/or too strange for enterprise sysadmins and became the main impetus for moving towards NFSRODS v1.0 with NFSv4 ACLs capable of expressing multi-ownership.

Other limitations of this permissions system include a lack of `chmod` and no workable mappings for groups. The projection of iRODS permissions out to the NFSv4.1 mountpoint works, but there is no way for a file owner to set permissions for others from within a mountpoint and be reflected back into the iRODS Catalog. As a preliminary release, group permissions were not mapped out to the mountpoint at all. Both of these will be addressed with NVSv4 ACLs in NFSRODS v1.0.

USAGE

Deployment of NFSRODS v0.8.0 requires some preparation and then three steps.

The preparation includes making sure that the necessary user UIDs and usernames are available for the different components (Enterprise VM, NFSRODS server, and within the iRODS Catalog). The three steps include configuration, the `docker run` command, and setting up the mountpoint.

Configuration

Configuration for NFSRODS includes three configuration files, two of which do not need changes from the distributed examples. The `exports` and `log4j.properties` files can be used as is.

The `server.json` file needs to be updated to point to the correct iRODS server:

```
{
    // This section defines options needed by the NFS server.
    "nfs_server": {
        // The port number within the container to listen for NFS requests.
        "port": 2049,

        // The path within iRODS that will represent the root collection.
        // We recommend setting this to the zone. Using the zone as the root
        // collection allows all clients to access shared collections and data
        // objects outside of their home collection.
        "irods_mount_point": "/tempZone",

        // The refresh time (in minutes) for cached user information.
        "user_information_refresh_time_in_minutes": 60,

        // The refresh time (in milliseconds) for cached stat information.
        "file_information_refresh_time_in_milliseconds": 1000
    },

    // This section defines the location of the iRODS server being presented
    // by NFSRODS. The NFSRODS server can only be configured to present a single zone.
    "irods_client": {
        "host": "hostname",
        "port": 1247,
        "zone": "tempZone",

        // Because NFS does not have any notion of iRODS, you must define the
        // target resource for new data objects.
        "default_resource": "demoResc"
    },

    // An administrative iRODS account is required to carry out each request.
    // The account specified here is used as a proxy to connect to the iRODS
    // server. iRODS will still apply policies based on the client's account,
    // not the proxy account.
    "irods_proxy_admin_account": {
        "username": "rods",
        "password": "rods"
    }
}
```

The `nfs_server` section of the configuration file defines the settings for the NFSv4 side of NFSRODS. This includes the `port` number to expose as NFS (default 2049), the `irods_mount_point` to define how deep within iRODS the mount-point will expose the virtual filesystem, and some cache settings (`user_information_refresh_time_in_minutes` and `file_information_refresh_time_in_milliseconds`) for how long the NFSRODS server will keep a local copy of information found from the underlying unix system or the iRODS catalog.

The `irods_client` section of the configuration file defines the settings for the iRODS client side of NFSRODS (`host`, port, and `zone`). The `default_resource` setting will define where any newly created files within the mountpoint are physically created within iRODS.

NFSRODS occasionally needs to take action within iRODS that it would not be able to take without a higher privilege level. In these cases, NFSRODS uses the proxy mechanism of iRODS to request actions on behalf of the requesting user. The `irods_proxy_admin_account` is used to configure a rodsadmin username and password.

Docker

Starting NFSRODS requires a single `docker run` command of the form:

```
$ docker run -d --name nfsrods \
          -p <public_port>:2049 \
          -v </full/path/to/nfsrods_config>:/nfsrods_config:ro \
          -v /etc/passwd:/etc/passwd:ro \
          -v /etc/shadow:/etc/shadow:ro \
          nfsrods
```

The options launch the image known as `nfsrods`, put the container into daemon mode, and define the name of the running container (`nfsrods`), the port mapping from the outside world into the container, the volume mount to the configuration files, and the volume mounts of the host system's `/etc/passwd` and `/etc/shadow` files.

Restarting the NFSRODS server will not affect existing mountpoints other than the requirement to re-fetch any lost cache information.

Mountpoint

Once the NFSRODS server is running, the standard `mount` command can be used to mount the remote filesystem and provide a location for regular users to get access to the iRODS namespace:

```
$ sudo mkdir <mount point>
$ sudo mount -o sec=sys,port=<public_port> <hostname>:/ <mount_point>
```

Note the `hostname` is the hostname where NFSRODS is running and the `:/` after the hostname express to the mount command to mount the entire namespace provided by NFSRODS.

If you do not receive any errors after mounting, then a unix user with a properly mapped UID and username should be able to access the mount point like so:

```
$ cd <mount_point>/path/to/collection_or_data_object
```

FUTURE WORK

NFSRODS v0.8.0 is a work in progress. It is known that the lossy nature of the mapping between the iRODS Permission Model and traditional Unix permissions is insufficient for enterprise usage.

In addition to the permissions model, NFSRODS needs a test suite that expresses its intended usage, reduced debug logging, as well as performance measurements to characterize its overhead.

A lossless mapping between iRODS and NFSv4 ACLs with multi-ownership has been developed and will be implemented by the end of 2019. It is expected that NFSRODS v1.0 will include the lossless mapping and be ready for production deployments.

SUMMARY

The demand for a virtual filesystem with included policy and well-understood semantics is very strong. iRODS provides that abstraction and capability. However, it takes a lot of engineering effort to teach existing tools and workflows to speak the iRODS protocol. It is more likely that tools can read and write into a mountpoint provided by a compatibility layer between POSIX and iRODS.

NFSRODS provides this compatibility layer and v0.8.0 is the first proof of concept. Existing tools can read and write into the iRODS namespace without any changes to their own code, and iRODS organizational policy is enforced on the server.

The lossy mapping between iRODS and traditional Unix permissions will be addressed in an upcoming v1.0 release where iRODS and NFSv4 ACLs will be losslessly mapped to one another.

REFERENCES

[1] iRODS Client NFSRODS. `https://github.com/irods/irods_client_nfsrods`

[2] Xu, H., Russell, T., Coposky, J., et al: iRODS Primer 2: Integrated Rule-Oriented Data System. In: Synthesis Lectures on Information Concepts, Retrieval, and Services. 131pp. Morgan Claypool. (2017)

[3] Haynes, T., Noveck, D.: Network File System (NFS) Version 4 Protocol (2015) `https://tools.ietf.org/html/rfc7530`

[4] NFS-RODS: A Tool for Accessing iRODS via the NFS Protocol (2016) `https://irods.org/uploads/2016/06/NFSRODS-slides.pdf`

[5] Callaghan, B., Pawlowski, B., Staubach, P.: NFS Version 3 Protocol Specification (1995) `https://tools.ietf.org/html/rfc1813`

[6] NFS4J. `https://github.com/dCache/nfs4j`

[7] Carl Boettiger, C.: An introduction to Docker for reproducible research (2015) `https://doi.org/10.1145/2723872.2723882`

[8] Kohl, J., Neuman, C.: The Kerberos Network Authentication Service (V5) (1993) `https://tools.ietf.org/html/rfc1510`

[9] Jargon - iRODS Java client library. `https://github.com/DICE-UNC/jargon`

[10] Traditional Unix permissions. `https://en.wikipedia.org/wiki/File_system_permissions#Traditional_Unix_permissions`

Integration of iRODS data workflows in an extensible HTTP REST API framework

Mattia D'Antonio
CINECA - Interuniversity
Consortium
Casalecchio di Reno (BO) -
Italy
m.dantonio@cineca.it

Claudio Cacciari
CINECA - Interuniversity
Consortium
Casalecchio di Reno (BO) –
Italy
c.cacciari@cineca.it

Giuseppa Muscianisi
CINECA - Interuniversity
Consortium
Casalecchio di Reno (BO) –
Italy
g.muscianisi@cineca.it

Michele Carpenè
CINECA - Interuniversity Consortium
Casalecchio di Reno (BO) - Italy
m.carpen@cineca.it

Giuseppe Fiameni
CINECA - Interuniversity Consortium
Casalecchio di Reno (BO) – Italy
g.fiameni@cineca.it

ABSTRACT

We developed a set of HTTP REST APIs on top of iRODS to support users of different communities to automate both ingestion and retrieval data workflows. We built a common REST APIs layer by implementing basic functionalities, including the interaction with iRODS, within an extensible framework (RAPyDO: Rest Apis with Python on Docker) that we developed and adopted to build communities-specific REST APIs. More in details, we are collaborating with the EUropean DATa infrastructure EUDAT Collaborative Data Infrastructure (CDI); European projects like EOSC-hub and SeaDataCloud; national initiatives in collaboration with Telethon Foundation (a non-profit organization for genetic diseases research) and SIGU (Italian Society for Human Genomics). All endpoints are written by using the Python language on the Flask framework. APIs are served through an uWSGI web server deployed within a Docker container. We created a wrapper of the python irods client (PRC) to let both the core framework and communities specific APIs for easily interact with iRODS by supporting all main authentication protocols like native passwords, Pluggable Authentication Modules (PAM), Grid Security Infrastructure (GSI). To be able to support all required authentication methods we contributed to the PRC development by implementing authentication modules for both GSI and PAM. Most of iRODS-based functions that we developed can be mapped against corresponding icommands like ils, iget, iput, imv, icp, imeta, irule, iticket but also more complex functionalities have been realized, for instance streamed read/write operations from/to network sockets. To be able to execute data intensive and complex workflows, we also introduced an asynchronous layer implemented on Celery, a task management queue based on distributed message passing.

Keywords

HTTP API, Python, data management, iRODS, REST API, authentication, PAM, GSI

iRODS UGM 2019, June 25-28, 2019, Utrecht, Netherlands

INTRODUCTION

Data management is becoming one of the most challenging topic in computer science, since requirements for data production and manipulation often overcame whom for data analysis. We at CINECA [1], the Italian Interuniversity Consortium, are involved in many European Projects and National Initiatives strongly based on the requirement of efficient and flexible data workflows. Every project has its own very specific set of constraints and requirements, but by comparing all of them, it is possible to identify several common needs. By working on these requirements, a shared data management layer can be implemented and used as a base for every data oriented project. In this paper, we will describe our current solution, based on a data management layer built over iRODS. A quick overview of the main projects that involve our group can highlight the common needs we are working on.

EUDAT Collaborative Data Infrastructure (CDI)

The EUDAT CDI [2] is a European e-infrastructure of integrated data services and resources to support research. This infrastructure and its services have been developed in close collaboration with over 50 research communities spanning across many different scientific disciplines and involved at all stages of the design process. The EUDAT CDI network provides a range of services for data upload, retrieval, identification, description, replication through a series of core services like B2SAFE [3] (data storage), B2STAGE [4] (data transfer) and many other (like B2HANDLE, B2SHARE, B2FIND, B2NOTE).

B2SAFE is a robust, safe and highly available service, which allows community and departmental repositories to implement data management policies on their research data across different geographical and administrative domains in a trustworthy manner. Currently iRODS is used at the EUDAT sites for the federation of the data nodes and the node-wise policy enforcement.

To offer functionalities for data transfer between EUDAT resources and external computational facilities we built the B2STAGE service. B2STAGE supports different functionalities allowing users to either stage data outside EUDAT or ingest computational results while maintaining the coherency of associated Persistent Identifiers (PIDs). This service offers two interfaces, one based on the GridFTP protocol [5] and one based RESTful HTTP endpoints. The main requirements of this service are the ease of use, the interoperability with other EUDAT services and the support for the automation of ingestion and retrieval workflows. The HTTP API interface of B2STAGE is built by adopting the common strategies discussed in this paper on top of B2SAFE and iRODS. Furthermore, B2STAGE HTTP-API is in turn the base to build other services (like SeaDataCloud HTTP API, see paragraph below) and is part of the EOSC-HUB.

EOSC-hub

The European Open Science Cloud (EOSC) [6] aims to offer open and seamless services for archiving, management, analysis and re-use of research data, across different scientific disciplines. Within this project the Hub, a single contact point for European researchers, represents a particular role and innovators to discover, access, use and reuse a broad spectrum of resources for advanced data-driven research. EOSC-hub brings together multiple service providers from the EGI Federation [7], EUDAT CDI, INDIGO-DataCloud [8] and other major European research infrastructures to deliver a common catalogue of research data, services and software for research. Through EUDAT CDI, B2STAGE is part of the EOSC-hub network, bringing to an increased level of flexibility and interoperability required from this service.

SeaDataNet - SeaDataCloud

SeaDataNet [9] is a standardized infrastructure for managing the large and diverse data sets collected by the oceanographic fleets and the automatic observation systems. This infrastructure aims to aggregate and standardize the highly fragmented marine observing system based on more than 600 scientific data collecting laboratories (from governmental organizations to private industries) in the countries bordering the European seas. Data are collected by using various sensors on board of research vessels, submarines, fixed and drifting platforms, airplanes and satellites, to measure physical, geophysical, geological, biological and chemical parameters, biological species etc. SeaDataNet infrastructure was implemented during the SeaDataNet project (2006-2011), grant agreement 026212, EU Sixth Framework Programme and consolidated in the second phase, SeaDataNet 2 project (2011-2015), grant agreement 283607, EU Seventh Framework Programme. SeaDataCloud project (2016-2020), grant agreement 730960, EU H2020 programme, is the third (and current) phase and it aims at considerably advancing SeaDataNet Services and increasing their usage, adopting cloud and High Performance Computing technology for better performance. This background highlights the main requirements that these services will have to satisfy: efficient and simple ingestion interfaces; support for data workflows for data manipulation, format conversion, quality verification; facilities for data searching and retrieval. SeaDataCloud (SDC) CDI HTTP API are built over B2STAGE HTTP API by extending the EUDAT services with the inclusion of custom endpoints. In particular, SDC introduced the integration with Rancher [10] for the execution of quality check workflows based on docker containers [11] and greatly enhanced the use of asynchronous endpoints (implemented by means of the Celery [12] framework).

The Genomic Repository

The Genomic Repository is a data and metadata archive born to meet the requirements from two different partners and leading to the definition of a single solution adopted for both the platforms. The Telethon Foundation [13], a non-profit organization for genetic diseases research, is involved in the Undiagnosed Diseases Program (UDP) [14] pursuing the goal of providing a diagnosis to pediatric patients with a genetic disease without a name. To match this goal all clinical and genomic data have to be stored into a common platform able to compare such information with similar databases across the world, looking for similarities and hopefully the identification of *second cases*. Genomic data, obtained from massive next-generation sequencing (NGS) platforms, require computationally intensive analyses available on HPC computational centers.

The second use case is from the NIG (Network of Italian Genome) project of the Italian Society for Human Genetics (SIGU) [14]. The main purpose of this project is the definition of an Italian Reference Genome for the identification of genes responsible for genetic diseases and susceptibility genes for complex diseases in both basic and translational researches; genetic variants responsible for inter-individual differences in drug response in the Italian population; new targets for both diagnosis and treatment of genetic diseases. This project required the creation of a shared database containing data from nucleic acids sequencing of hundreds to thousands of Italian subjects. Sequencing data are then analyzed by executing HPC bioinformatics workflows and results are stored into the database and linked to phenotypic information. A final procedure collects all data to produce aggregated results such as distinctive Italian variants, variant frequencies, variant geographical distribution, and genotyping distribution.

Both the projects required to access to information in an automatable way to be able to interact with external initiatives and the opportunity to share data on HPC clusters in a secure way. These requirements have been achieved with the adoption of HTTP APIs and iRODS. In particular, iRODS has been identified as an important technology to make data available on computing nodes by preserving the high security levels needed to treat with human clinical data.

THE RAPYDO FRAMEWORK

To share technological solutions in very different contexts, from oceanography to genomics, we implemented a common framework adopted by all these projects and named RAPyDo [15] (acronym for Rest Api with Python on Docker). RAPyDo is a wrapper for docker-compose [16], a tool for defining and running multi-container Docker [17] applications. Docker is a tool implementing high-level APIs to provide lightweight containers that run processes in isolated environments by packaging an application and its dependencies in a virtual container. RAPyDO is able to merge several levels of configurations (base configurations, project-level configurations and deployment-level configurations) to create dynamic docker-compose definitions allowing easy deployment on every platform (from its own laptop for development purpose to production servers). Furthermore, RAPyDO is an extensible and modular framework and it implements basic functionalities and services integrations that can be used or extended by projects. RAPyDo supports iRODS as a base technology for data management and implements a wrapper client based on the python-irods-client [18]. RAPyDO projects can be administrated and deployed through a controller script implementing all the logics over docker-compose and able to merge configurations (e.g. to enabled and disabled supported services) and functionalities (e.g. base and custom endpoints). An overview of the main components is described below.

RAPyDo Architecture

RAPyDO is mostly implemented in Python programming language, in particular the controller and the http-api components. An optional Web interface is implemented with Angular.

HTTP-APIs are powered by python Flask micro-framework and provides access to set of RESTful HTTP API connected to all the other backend services (e.g. database, tasks queue, data storage, etc). When development mode is enabled, REST endpoints are directly served by Flask through uWSGI, in production mode an nginx reverse proxy is deployed ahead of the backend to enhance security and introduce SSL certificates management. Supported services, when enabled, are automatically deployed as docker containers but can be configured as external resources with independent deployment. An overview of the main components is show in Figure 1

Supported services

Databases

RAPyDo supports several databases and in particular relational databases (MySQL, MariaDB, PostgreSQL, SQLite), graph databases (Neo4j) and no-sql databases (MongoDB). All these databases can be enable to support the authentication functionalities to manage usernames, password, groups and sessions (based on JWT tokens)

Celery

HTTP-APIs endpoints typically implement synchronous operations by directly providing to the client the result of the requested resources. This basic schema implies fast responses (no more than few seconds) to avoid connection timeouts but this not always can be achieved, in particular when handling with huge amount of data. Asynchronous endpoints provide to the client an operation identifiers than can be used to track the progress of the request until it completes. RAPyDo adopted the Celery framwork to introduce asynchronous operations. Celery is a task management queue based on distributed message passing and built on top of RabbitMQ.

iRODS

RAPyDo adopted iRODS as base layer for data management, since this technology offers several valuable advantages. iRODS implements data virtualization, allowing access to distributed storage assets under a unified name-space and allows to access from multiple clients: command line (icommands), web interface (webdav), ftp transfer (grid-ftp). Furthermore, it integrates certificate-based authentication with the GSI plugin, irules to trigger actions based on events at collection or data objects level, access control list (ACL) mechanisms to allow for the implementation of complex access rights policies preserved at every level, regardless of the access method.

The different access methods and the support for ACLs allow to successfully integrating the software stack with the use High Performance Clusters (HPC) for analysis purpose. In particular all data stored into iRODS can be shared on the different components by preserving all the security policies.

Rancher

The use of High Performance Clusters is not always possible or flexible enough. In specific contexts, Docker can be used to execute analyses, quality checks, data validation tasks and other project-specific software. Rancher is an open source multi cluster management platform used to deploy and manage docker containers (through the Cattle engine up to version 1.6 and through Kubernetes from version 2.0). RAPyDO HTTP APIs are able to communicate with Rancher APIs (version 1.6) to automatically deploy and interact with docker containers executed in a distributed environments.

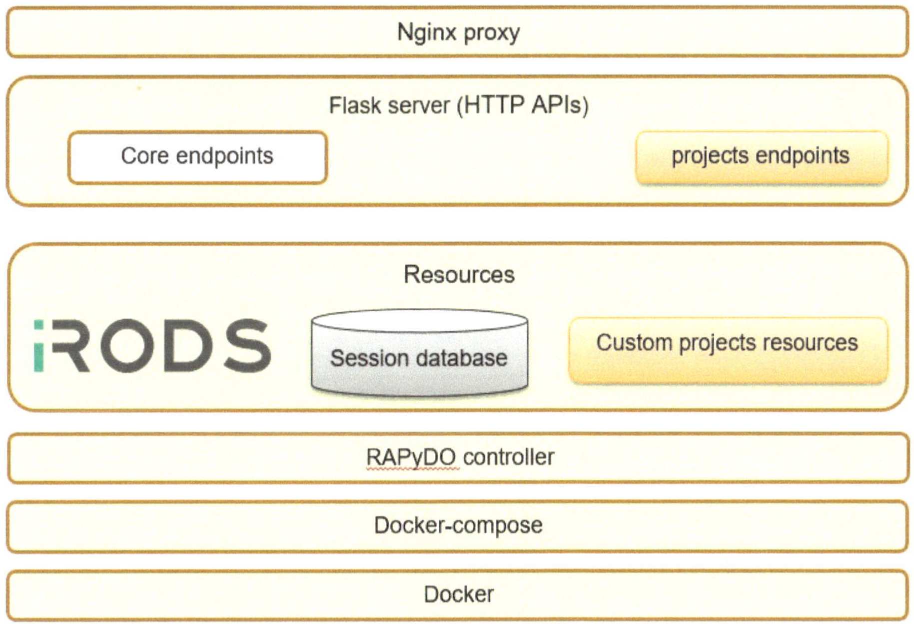

Figure 1. Architecture of the main components of the RAPyDo framwork

IRODS CLIENT

iRODS officially supports Python through the python-irods-client [18] (PRC), an open source project released on Github. RAPyDO HTTP APIs integrates PRC to implement all iRODS-based operations on both Flask (HTTP sync endpoints) and Celery (asynchronous tasks). We created a wrapper client based on the python-irods-client to implement utility operations used from both endpoints and tasks. Most of the methods implemented by this client can be divided in three main categories:

- Methods that can be strictly mapped on icommands, e.g. list(), mkdir(), copy(), put(), get(), move(), remove(), set_permissions(), get_metadata(), ticket() and others. These methods can be respectively mapped on ils, imkdir, icopy, iput, iget, imv, irm, ichmod, imeta ls, iticket.

- Simple utilities methods without a corresponding icommand, e.g. exists(), is_collection(), is_dataobject() and others
- Method to perform more complex operations, e.g.
    - write_file_content(path, string)
    - get_file_content(path) returning a string
    - read_in_chunks(path, chunk_size) returning an iterator
    - read_in_streaming(path) directly streaming data object content as Flask stream
    - write_in_streaming(path) directly writing data object from Flask streams

This list not exhaustive since it only include general purposes methods, took as examples.

Authentication

RAPyDO HTTP APIs support all main iRODS authentication protocols and in particular native credentials, Grid Security Infrastructure (GSI) and Pluggable authentication modules (PAM).

Native credentials (usernames and passwords defined into the iRODS database) is the default authentication protocol and the python-irods-client natively supports it. In a production environment, iRODS is often secured by means of GSI certificates or by using heterogenes methods (e.g. LDAP) supported by PAM protocol. Since PRC was lacking the support for GSI and PAM authentication, we contributed to the development by implementing the specific missing modules. The GSI integration is a completed task and the contribution is merged into the main branch since around January 2017. The PAM integration is completed respect to some use cases but requires improvements to achieve a larger audience. The current PAM module is merged into the main branch since December 2018.

To implement the Pluggable authentication modules (PAM) support we introduced a PluginAuthMessage to encapsulate PAM requests

```
class PluginAuthMessage(Message):

    _name = 'authPlugReqInp_PI'

    auth_scheme_ = StringProperty()

    context_ = StringProperty()
```

The same class is also used to perform GSI requests.

PAM requests are propagated to the iRODS server by sending ah authorization request with type RODS_API_REQ and apiNumber 1201, equivalent to AUTH_PLUG_REQ_AN. The body of the message is a PluginAuthMessages containing a context based on user, pam password and ttl.

```
ctx_user = '%s=%s' % (AUTH_USER_KEY, self.account.client_user)

ctx_pwd = '%s=%s' % (AUTH_PWD_KEY, self.account.password)

ctx_ttl = '%s=%s' % (AUTH_TTL_KEY, "60")

ctx = ";".join([ctx_user, ctx_pwd, ctx_ttl])

message_body = PluginAuthMessage(

    auth_scheme_=PAM_AUTH_SCHEME,

    context_=ctx

)
```

```
auth_req = iRODSMessage(

    msg_type='RODS_API_REQ',

    msg=message_body,

    int_info=1201

)
```

The server response in stored in a AuthPluginOut:

```
class AuthPluginOut(Message):

    _name = 'authPlugReqOut_PI'

    result_ = StringProperty()
```

Containing a new temporary password with validity equivalent to the negotiated ttl and used to initialize a native password connection

```
auth_out = output_message.get_main_message(AuthPluginOut)

self._login_native(password=auth_out.result_)
```

To implement the Grid Security Infrastructure (GSI) a more complex workflow has been introduced. As high-level overview, the GSI protocol can be summarized in three main steps:

1) send to iRODS server a message to request GSI authentication. This step is similar to the PAM use case but providing a different auth_scheme (GSI_AUTH_PLUGIN) and a different context (AUTH_USER_KEY=self.account.client_user)
2) create a context handshaking GSI credentials by creating a GSI SecurityContext to send the GSI client token and to receive corresponding token from the server
3) Complete the protocol by sending the username through a requesto to the iRODS server with an apiNumber 704, equivalent to AUTH_RESPONSE_AN

CONCLUSIONS

In this paper, we presented a modular and extensible framework (RAPyDo) that we developed to share common strategies among European and National projects whom we are involved into. Our main intention is to extract from each project all requirements that can be satisfied by applying previously identified and implemented solutions. New requirements are analyzed to identify candidates for generalizable solutions to be included in the framework and to be reused in further projects. In this way the framework can growth and enhance itself at each application. RAPyDo supports several services (databases, task queues, interfaces) and in particular adopted iRODS as basic data management technology. The HTTP APIs module included in RAPyDo also introduces a python client based on the official python-irods-client (PRC) implementing utility functions and authentication protocols like GSI and PAM. We already adopted this framework in several European and National projects like the EUDAT Collaborative Data Infrastructure (CDI), the Hub of European Open Science Cloud (EOSC-hub), SeaDataCloud HTTP APIs and a Genomic Repository built in collaboration with Telethon Foundation and Italian Society for Human Genetics (SIGU). All these projects are data repository based on iRODS technology. Furthermore, the framework is also adopted onto European Projects not based on iRODS, like I-Media Cities (grant agreement N° 693559E from the uropean Union's Horizon 2020 research and innovation programme) and Meteo Italian Supercomputing Portal MISTRAL (funded under the Connecting Europe Facility (CEF) – Telecommunication Sector Programme of the European Union). As future perspectives, we plan to enhance the current implementation by expanding the number of supported projects. Furthermore, we need to fight against the natural obsolescence of adopted solutions by continuously improve the framework by investigating state-of-the art technologies candidate to be included in the supported stack.

ACKNOWLEDGMENTS

The implementation of the GSI and PAM modules started by forking the code of the python irods client [18] and PRC is also a base component of our APIs.

For the implementation of the GSI plugin a special mention is for Roberto Mucci and Paolo D'Onorio De Meo, not included as authors of this paper. Paolo D'Onorio De Meo is also one of the main developers of the RAPyDo framework.

EUDAT is funded by from the European Union's Horizon 2020 research and innovation programme under grant agreement No. 654065.

EOSC-hub is funded by from the European Union's Horizon 2020 research and innovation programme under grant agreement No. 777536

SeaDataCloud is funded by from the European Union's Horizon 2020 research and innovation programme under grant agreement No. 730960

RAPyDo is not strictly bound to any project and it is publicly released as open source code on github [15] under MIT License. Permission is hereby granted, free of charge, to any person obtaining a copy of the software and associated documentation files to deal in the Software without restriction, including the rights to use, copy, modify, merge, publish, distribute, sublicense, and/or sell copies of the Software.

REFERENCES

[1] CINECA - Italian Interuniversity Consortium, https://www.cineca.it/en

[2] EUDAT CDI, https://eudat.eu/eudat-cdi

[3] B2SAFE, https://www.eudat.eu/b2safe

[4] B2STAGE, https://www.eudat.eu/b2stage

[5] GridFTP - Globus Toolkit, http://toolkit.globus.org/toolkit/docs/latest-stable/gridftp/

[6] EOSC-hub, https://www.eosc-hub.eu

[7] EGI Federation, https://www.egi.eu/federation/

[8] INDIGO DataCloud, https://www.indigo-datacloud.eu/

[9] SeaDataNet, https://www.seadatanet.org/

[10] Rancher: Container Orchestration, https://rancher.com/

[11] Docker Container Platform, https://www.docker.com/

[12] Celery: Distributed Task Queue, http://www.celeryproject.org/

[13] Telethon Foundation, http://www.telethon.it/en

[14] SIGU - Italian Society for Human Genetics, https://www.sigu.net/

[15] RAPyDo Framework, https://github.com/rapydo

[16] Docker Compose, https://docs.docker.com/compose/

[17] Docker Container Platform, https://www.docker.com/

[18] Python iRODS Client (PRC), https://github.com/irods/python-irodsclient

Rodinaut: A tool for metadata management

Othmar Weber
Bayer
Leverkusen, Germany
othmar.weber@bayer.com

ABSTRACT

For scientists who need to ensure compliance with data security/privacy, and find information in iRODS, Rodinaut is a web application that enables viewing and managing metadata. Unlike the existing command-line tool, our product is self-explaining, easy and fast to use, and improves user experience with iRODS.

iRODS UGM 2019, June 25-28, 2019, Utrecht, Netherlands

More than Just Load Balancing iRODS Using HAProxy

Tony Edgin
CyVerse, University of Arizona
Tucson, Arizona
tedgin@cyverse.org

ABSTRACT

The iRODS community has demonstrated that HAProxy works well for horizontally scaling iRODS catalog providers, but HAProxy can provide other functionality. This talk will demonstrate how HAProxy can access information about the user and client application and use it to control quality of service through throttling per user access, providing fast lanes for certain applications, and routing data transfer sessions to specific catalog providers. The talk will also show how HAProxy can be configured to filter out port scanners. Finally, the talk will show how to configure HAProxy to allow a single canonical host name for iRODS and other services like davrods.

iRODS UGM 2019, June 25-28, 2019, Utrecht, Netherlands

Migrating data when decommissioning PetaBytes of storage

John Constable
Wellcome Sanger Institute
Hinxton, England
jc18@sanger.ac.uk

ABSTRACT

Wellcome Sanger Institute has ~18PB of genomic data in 399 resources on 76 resource servers across six Zones. This is the story of what happened when we needed to retire some of the servers.

iRODS UGM 2019, June 25-28, 2019, Utrecht, Netherlands

Surgical Critical Care Initiative (SC2i): Leveraging iRODS to Accomplish Multi-Site Data Collection, Harmonization, and Analytics to Generate Clinical Decision Support Tools

Andy MacKelfresh
Duke University
Durham, NC
andy.mackelfresh@duke.edu

Justin James
Renaissance Computing Institute (RENCI)
UNC-Chapel Hill
jjames@renci.org

ABSTRACT

To support the development of Clinical Decision Support Tools (CDSTs) in both civilian and military health systems, the Surgical Critical Care Initiative (SC2i), a Department of Defense funded consortium of Federal and non-Federal institutions, leverages iRODS to harmonize clinical, laboratory, and bio-bank data and centralize 30+ million high-quality data elements to enhance complex decision making in acute and trauma care.

iRODS UGM 2019, June 25-28, 2019, Utrecht, Netherlands

iRODS S3 Resource Plugin: Cacheless and Detached Mode

Justin James
Renaissance Computing
Institute (RENCI)
UNC Chapel Hill
jjames@renci.org

Terrell Russell
Renaissance Computing
Institute (RENCI)
UNC Chapel Hill
unc@terrellrussell.com

Jason Coposky
Renaissance Computing
Institute (RENCI)
UNC Chapel Hill
jasonc@renci.org

ABSTRACT

The iRODS S3 Resource Plugin has been updated with a new set of operating modes. It can now be configured to operate without the need for a local replica to be housed in a cache resource under a compound resource. Additionally, in detached mode, iRODS will not need to redirect to a particular iRODS server before connecting to the S3 object store. This functionality is made available alongside iRODS 4.2.6.

Keywords

iRODS, S3, AWS, cacheless, data management

INTRODUCTION

iRODS has been capable of surfacing S3-compatible object stores for more than a decade. Amazon launched their Simple Storage Service (S3) in March of 2006[1]. The original iRODS file driver code for S3 was written in 2009 and included in iRODS 2.2[2]. The iRODS S3 Resource Plugin was pulled from the main code and ported to the plugin architecture in 2013 with E-iRODS 3.0[3].

Since its inception, the iRODS S3 interface depended on being surfaced through a compound resource that held both cache and archive resources. The cache resource would hold 'local' replicas of data and be available for POSIX operations. The archive resource would hold non-POSIX interface replicas (object stores, tape formats, etc.) and usually require access with greater latency.

```
$ ilsresc
s3compound:compound
|-- s3archive:s3
'-- s3cache:unixfilesystem
```

The S3 plugin itself only implemented a few operations:

```
irods::RESOURCE_OP_UNLINK
irods::RESOURCE_OP_STAT
irods::RESOURCE_OP_RENAME
irods::RESOURCE_OP_STAGETOCACHE
irods::RESOURCE_OP_SYNCTOARCH
```

All of the other POSIX-style operations were handled by the cache resource.

iRODS UGM 2019 June 25-28, 2019, Utrecht, Netherlands
[Authors retain copyright.]

This compound configuration was flexible but required a cache management policy to be devised, implemented, and executed unique to each deployment. This burden has proven more than most deployments wish to maintain. There has been a stated desire for an iRODS Consortium-backed cache management policy or set of rules to help with the 'normal' cases. Since every deployment has different opinions and requirements about what proper cache management looks like, a more elegant solution has presented itself... not having a cache to manage in the first place.

Network speeds have increased. Affordable access and availability to the cloud has increased. Both the academic and commercial comfort level with data in the cloud has increased. It was time for the S3 resource to manage its own temporary cache and not require additional complexity to perform well.

This initial release of the iRODS S3 Resource Plugin with Cacheless and Detached Mode satisfies this design goal and provides a simpler way to attach an S3-compatible storage system to iRODS.

MODES

The new plugin now supports three operating modes. This mode is set using the `HOST_MODE` parameter in the resource context string. If the `HOST_MODE` is not set, the default is `archive_attached`, which operates the same way as the legacy S3 plugin.

	Archive	Cacheless
Attached	archive_attached (default)	cacheless_attached
Detached	N/A	cacheless_detached

Table 1. Three modes for the iRODS S3 Resource Plugin >= 2.6.1

Note that "archive_detached" is not a valid entry and will be ignored.

The columns of Table 1 represent the behavior of the plugin. 'Archive' mode means the S3 resource acts in the archive role behind or under a compound resource. It requires a cache resource to provide POSIX semantics and it must be attached to a specific iRODS server configured with the S3 credentials. 'Cacheless' mode means the S3 resource can be standalone and may be detached from any specific iRODS server (as per 'attached' or 'detached'). In this role, the S3 plugin provides POSIX semantics with no cache resource and requires no explicit cache management policy.

The rows represent the configuration determining which iRODS server will answer a particular S3 request. 'Attached' means that only the server that is defined as the host in the resource configuration will serve the request. 'Detached' means all iRODS servers may serve a request for a data object. This is appropriate if all servers have connectivity to the S3 backend and are designed to provide a horizontal scale-out solution.

ARCHITECTURE

To implement the cacheless S3 resource, we implemented the missing operations that the legacy 'archive-only' S3 resource had not implemented.

These include:


```
irods::RESOURCE_OP_OPEN
irods::RESOURCE_OP_READ
irods::RESOURCE_OP_WRITE
irods::RESOURCE_OP_CLOSE
irods::RESOURCE_OP_LSEEK
```

S3FS

The prototype implementation started with S3FS[4], an open source C++ FUSE-based filesystem presentation of S3.

The first step was to validate that S3FS was stable. An S3FS mount point was created and then an iRODS resource was configured with its vault assigned a directory within the mount point. A suite of iRODS tests was run and all the tests passed, except for a few bundle test cases. This was convincing enough that S3FS was a good starting point for the cacheless plugin.

The next step was to translate the FUSE operations to iRODS resource plugin operations. The iRODS resource plugin operations follow POSIX semantics instead of FUSE semantics. In order to implement this, additional state information about every open file must be stored and maintained.

The cacheless S3 plugin keeps a map of open file descriptors and file offsets to objects in S3. Any time the `RE-SOURCE_OP_OPEN` operation is called, a unique file descriptor is generated and a new entry in this map is created with an offset of 0. This file descriptor is the same file descriptor that is known to the iRODS core via the file first class object.

As reads, writes, and seeks are performed, the offset in the map is updated accordingly to match the POSIX behavior. If a file is opened multiple times (for example in a parallel read or write) each opened file is given a unique file descriptor and offset. When the `RESOURCE_OP_CLOSE` operation is called the appropriate entry is deleted from the map.

Mitigating Global Variables

S3FS uses many global variables for S3 configuration, internal data structures, flow control, etc.

These can conflict when using parallel uploads and downloads or when writing to more than one resource within the same agent (e.g. when two S3 resources are children of a replication resource in a resource hierarchy).

Care was taken to mitigate this by 1) using "thread_local" variables so that each thread manages its own values and 2) storing configuration information in the `irods::plugin_property_map` making the iRODS catalog the source of persistent truth.

Two Protocols

The cacheless S3 plugin must provide a live translation of the conversation between two different protocols, single buffer or parallel iRODS on one side and multipart S3 on the other. For downloads from S3, this is straightforward since multipart S3 data is translated and sent to the iRODS client. For uploads, the plugin handles this translation by writing chunks of data to a temporary on-disk scratch space and then reading them back out before cleaning the scratch space. Overall, the plugin provides a management-free, cacheless experience to the iRODS administrator.

PERFORMANCE

Included in this section are some basic assessments of the plugin's performance and the steps taken to improve that performance. Both download and upload performance are compared against the AWS S3 CLI.

Download

Problem

When iRODS performs large file / parallel downloads, the plugin receives requests for bytes in a seemingly random order. If these individual random download requests are performed separately and in an uncoordinated manner, the S3 multipart download performance is not optimized.

Additionally, S3FS core code does read-ahead and will retrieve more than is requested which helps download performance but was still significantly slower than using the AWS CLI.

Goals

The problems above stood in the way of two goals for download performance. First, for this cacheless S3 plugin to be a viable alternative for users, download performance must be reasonably close to the performance of using the AWS CLI. Second, this plugin must be able to quickly service small requests within large files. If a file is 100GB in size, but a request is only for 1K of its data, this plugin should definitely not download the entire file.

Solution

The following approach solved both problems.

First, only read the requested bytes when requested, do not read ahead. Designate that if a second subsequent simultaneous read is requested this means the client is interested in performing a parallel download. If this occurs, initiate a full multipart S3 download of the requested file. Then, all open threads wait for a notification that a multipart chunk has been downloaded. On each multipart completion, the threads check if their requested bytes have been downloaded. If so, return these bytes. If not, wait for the next notification.

Results

Downloads times with the cacheless S3 plugin are very close to download times using the AWS CLI using the same `S3_MPU_CHUNK` size and `S3_MPU_THREAD` count. If a user only requests a small part of a large file, this is returned quickly. No full file downloads are performed unless requested.

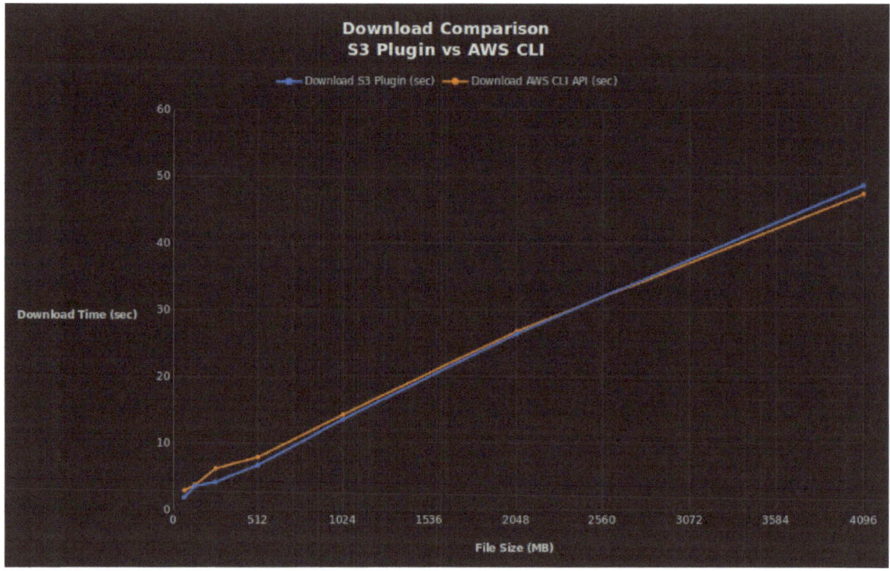

Figure 1. Download performance is nearly identical to AWS CLI.

	64MB	128MB	256MB	512MB	1024MB	2048MB	4096MB
Cacheless S3 Plugin	1.77	3.52	4.10	6.63	13.56	26.35	48.65
AWS S3 CLI	2.84	3.70	6.14	7.83	14.28	26.73	47.36

Table 2. Download times in seconds (average over 5x runs)

Upload

Problem

When uploading files through iRODS to S3, the full file must be uploaded. For large files, iRODS sends multiple data buffers in parallel. A naive approach would be to write these buffers directly to S3 but this requires rewriting parts of the file multiple times as S3 is an object store and can only write the entire object at a time.

Solution

While the S3 object is opened for writes, the data buffers are written to a local scratch file on the iRODS server. When the last `close()` is performed, flush the scratch file to S3 and remove it.

Results

File uploads are roughly 50% slower using the S3 plugin than using the AWS CLI but significantly better than the naive approach. The delay of writing and then reading the entire scratch file explains the increased linear difference in upload speed.

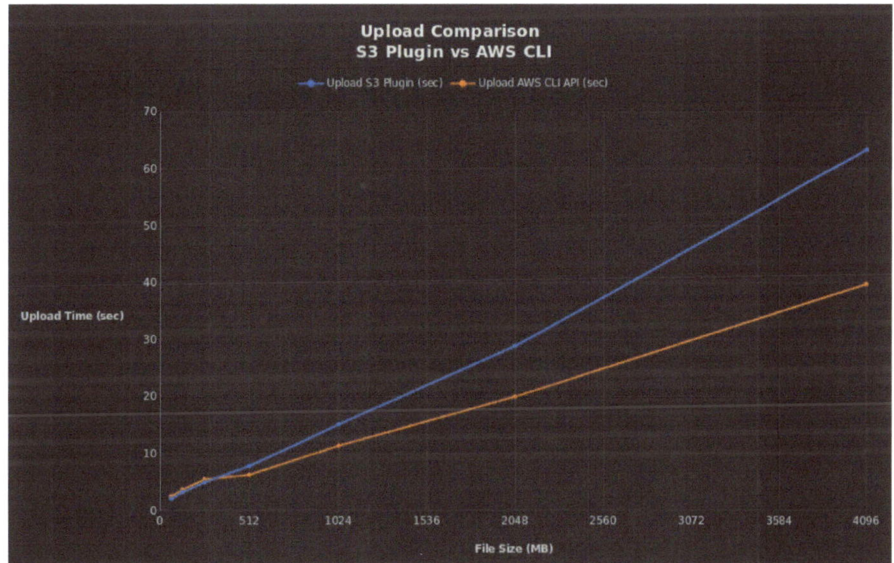

Figure 2. Upload performance loses ground with larger files... will investigate

	64MB	128MB	256MB	512MB	1024MB	2048MB	4096MB
Cacheless S3 Plugin	2.08	3.24	4.97	7.80	15.14	28.87	63.34
AWS S3 CLI	2.46	3.62	5.44	6.24	11.38	19.99	39.77

Table 3. Upload times in seconds (average over 5x runs)

USAGE

Using the cacheless S3 plugin is seamless. There is nothing to configure other than the mode itself in the context string.

The additional key/value pair in the context string could be `HOST_MODE=cacheless_attached`

Configuration

The following single `iadmin mkresc` command creates a new s3 resource of the name `news3resc` in `cacheless_attached` mode. The `bucket_name` specifies the S3 bucket. The S3 `prefix/to/iRODS/Vault` is optional. The backslashes at the end of each line are presentational only to ensure readability in this PDF.

```
iadmin mkresc news3resc s3 \
$(hostname):/bucket_name/prefix/to/iRODS/Vault \ "S3_DEFAULT_HOSTNAME=s3.amazonaws.com;\
S3_AUTH_FILE=/var/lib/irods/news3resc.keypair;\
S3_REGIONNAME=us-east-1;S3_RETRY_COUNT=1;\
S3_WAIT_TIME_SEC=3;S3_PROTO=HTTP;\
ARCHIVE_NAMING_POLICY=consistent;\
HOST_MODE=cacheless_attached"
```

Demonstration

With the cacheless S3 plugin properly configured, confirming correct behavior is straightforward. There is no second replica and the file sent is the same as the file returned:

```
$ ipwd
/tempZone/home/rods

$ iput -R news3resc foo

$ ils -L foo
  rods              0 news3resc          234 2019-06-25.14:40 & foo
        generic     /bucket_name/prefix/to/iRODS/Vault/home/rods/foo

$ iget foo bar

$ diff foo bar

$ echo $?
0
```

FUTURE WORK

The cacheless S3 plugin has passed all CI tests and has been released to the community. There are a few open issues, but no known configuration bugs at this point.

There are plans to focus on and improve the upload performance to try and match the AWS CLI performance as well as additionally implement the `RESOURCE_OP_READDIR` operation to facilitate recursive registrations.

The `S3_DEFAULT_HOST` will be improved to handle a comma separated list.

One of the most interesting requested enhancements is for the S3 plugin to retrieve its S3 credentials from the catalog or some other service like Hashicorp's Vault[5].

The iRODS Consortium has plans to take what has been learned with the S3 plugin and implement cacheless versions of the other iRODS archive resource plugins (WOS, etc.).

SUMMARY

The work to get the cacheless S3 plugin ready for release was more than expected. It presented some tricky scenarios and global variables which needed to be refactored. But it works and the performance is comparable to the AWS S3 CLI for downloads. Uploads still need some work.

The iRODS Consortium plans to continue this line of work with other plugins and remove configuration and maintenance complexity for iRODS administrators.

REFERENCES

[1] Amazon S3 (2006) `https://en.wikipedia.org/wiki/Amazon_S3`

[2] Wan, Mike: Initial S3 File Driver commit (2009).
`https://github.com/irods/irods-legacy/commit/2d204c14687340828483abecf8f73a8ea4dea944`

[3] E-iRODS 3.0 (2013) `https://github.com/irods/irods/tree/f67864a8d89251fc9288f6dc43f6c21191f81af1`

[4] s3fs-fuse: FUSE-based file system backed by Amazon S3 `https://github.com/s3fs-fuse/s3fs-fuse`

[5] Vault by Hashicorp `https://www.vaultproject.io`